MIND *and*
MATTER

MIND *and* MATTER

A LIFE IN MATH AND FOOTBALL

John Urschel

and Louisa Thomas

Penguin Press
New York | 2019

PENGUIN PRESS

An imprint of Penguin Random House LLC

penguinrandomhouse.com

ISBN 9780735224865 (hardcover)
ISBN 9780735224872 (ebook)

Printed in the United States of America
1   3   5   7   9   10   8   6   4   2

BOOK DESIGN BY SABRINA BOWERS

For Joanna

Contents

Preface

I am a mathematician, a PhD candidate at MIT.

I am also a former professional football player, a retired offensive lineman for the Baltimore Ravens.

Many people see me as a walking contradiction. They think that the pursuit of excellence in football makes the pursuit of excellence in mathematics impossible. They think that a strong interest in one makes a strong interest in the other improbable. People tend to think in binaries. Right and wrong. Black and white. Mind and body. Math and football.

I don't spend a lot of time wondering about the ways in which I'm an anomaly. My life is the only one I know. It's normal to me. We all have multiple and sometimes diverging identities. In different ways, math and football are both essential to me.

I'm sometimes asked about the connection between math and football. People want to know what edge being good in a classroom gave me on the field. I know what they want to hear, and I usually give it to them. I talk about basic physics, intelligence, and problem solving. But the truth is, football and math are disjoint in my experience. When the ball is snapped, I'm not thinking about vectors and forces. I'm not really thinking about much of anything. I'm simply moving.

Math gives me a way of making sense of the world. It helps me see past the confusion of everyday life and glimpse the underlying structures of the universe. It reveals the properties of shapes and the prevalence of patterns. It describes the relationships between things. I'm drawn to the rigor and clarity of mathematics, and to the elegance and simplicity of solutions to even the most complex problems. I know no other feeling like the satisfaction of discovery. I still feel the same sense of wonder and curiosity when I try to prove a theorem that I felt when I did logic and math puzzles as a kid.

Football put me in contact with something messier, something elemental and deep within me. It strengthened not only my body, but also my confidence and will. Football was a game, and I had a lot of fun playing it. But it was always also a test for me. I look like a football player: I am 6-foot-3, with shoulders that span a doorway, and massive hands and feet. But I never had as much raw athletic talent as a lot of the guys I played with and against. I relied on my intensity and competitiveness and desire.

THIS MEMOIR PROCEEDS on alternating parallel tracks. One is the story of my life in mathematics. It begins when I was a toddler, the age that my mother started buying me puzzle books and workbooks and started playing games with me, encouraging my mathematical creativity and reasoning. If I am truly an outlier, it is because of her—an

African-American single mother who loved math but was discouraged from it, who wanted me never to feel that any door was closed to me. For me, math is the great intellectual pursuit—though it calls upon my spirit too. It is a story of moving between the ideal, abstract world and the reality we live in, a story of private investigation and also collaboration—both with mentors such as Vadim Kaloshin, who first introduced me to mathematical research, and Ludmil Zikatanov, my collaborator and close friend, and with figures from the past: Leonhard Euler, Henri Poincaré, John von Neumann, and others, whose work my own has built on. It explores not only my experience of encountering and learning mathematical concepts, from infinity to graph theory to the uncertainty principle, but also learning how to think.

The other story is of my life in football, from the time I saw a picture of my father in pads. It is the story of how I strengthened my body and my will, how I learned to fail and learned from failing. It focuses on my experiences on the field, but it is also the story of how I learned to make friends and find my place in the social world. My football career began a decade or so after I saw that picture, at Canisius High School in Buffalo, New York, and continued at Penn State, during years when the team's very existence was threatened by scandal. Finally, I made my way to the Baltimore Ravens, in the NFL, where I learned to navigate a different kind of landscape, and faced a final difficult test: determining when to walk away.

Those two stories, as different as they are, converge in me.

MIND *and* MATTER

‖‖

One

Puzzles

Math, 1991–2003

When I was a toddler, my mother struggled to put things out of my reach. She would hide birthday and Christmas presents in obscure places around our house, but I would find them. If there were cookies in the kitchen, I would unlatch the dishwasher door and step on it to reach the counter. She tried to set rules, but I slipped around them. I didn't like being told what I could and couldn't do.

My mother is tall, nearly six feet, with dark brown skin, broad shoulders, and a confident smile. She is smart and stubborn, but she met her match in me. I was a troublemaker. From my perspective, though, I wasn't causing problems. I was solving them. When I wanted something, I would learn how to get it.

She figured out how to distract me. Even before I started school, she sat me at the kitchen table and put a workbook in front of me. She had me trace letters and numbers, and started to teach me how to

1

read. She did not stop buying the workbooks even after I started kindergarten. There were piles of them around the house and in the car, so that I would always have something to do. My mother, very wisely, gave me no indication that they had very much to do with what I was supposed to be learning at school.

Education was important to her, just as it had been important to her parents. Her own mother had grown up in segregated South Carolina. Her school occupied a room in the basement of the local black church. One teacher taught all the students, no matter their age. School ended in the eighth grade. My grandfather was from Alabama. As the oldest sibling, he had to leave high school in order to make money to support his family, which he did by paving roads. He eventually got his high school equivalency diploma, but not until he was twenty-five. Both of them moved to Cincinnati, where they got blue-collar jobs. My grandmother became a seamstress. My grandfather worked two jobs: the morning shift at a linen and uniform supply company and the afternoon shift at the General Motors factory.

My mother grew up in a rough part of Cincinnati. She attended public schools where the goal was just to graduate, not to expand a student's mind. Her high school had high dropout and teen-pregnancy rates. Fortunately, her math teachers recognized that she had an unusual aptitude and placed her in classes above her grade level. By the time she entered her senior year, she had taken all of the math classes her school had to offer. One of her teachers enrolled her in a college calculus class and arranged for the school to pay her tuition. She was valedictorian of her graduating class. Even so, her guidance counselor encouraged her to become a secretary. She didn't listen. She got a full scholarship to the University of Cincinnati. She became a nurse, and worked as a nurse at night in the operating room while putting herself through graduate school to earn a master's degree in biomedical science. While working as a nurse in Cincinnati, she met my father, a surgeon, and they moved to Edmonton, Alberta, and then to Winnipeg, where I was born in 1991. When they moved to Buffalo, she got her law degree. My parents separated when I was three and then divorced, and my father moved back across the border to Canada. My mother started working as a lawyer. As an African-American single mother, she knew all too well that life was hard. I couldn't count on easy opportunities. But she was determined that I would have an easier time than she did. She was a woman of great faith—in God, in education, and in me.

She was determined to encourage my education, but she was artful about it, not overbearing. She let me take the lead. My mother loved reading, and our house in Buffalo was filled with books of every kind: Fyodor Dostoyevsky, Anne Rice, Winston Churchill biographies, Bram Stoker. (She loved *Dracula* especially.)

I might find her reading cookbooks or about Thomas Jefferson and the pirates of Tripoli. But when it became clear that those kinds of stories did not really capture my imagination, she did not push me toward writing or history. I devoured the math and science books, and so she encouraged me to do more. We would sit at the table together, and she would watch me work with large, thoughtful dark eyes. Then suddenly her face would break into the widest smile. She liked to have fun. She made math a game, which, in some ways, it is and always has been.

Since I was her child, she was convinced that I was a little genius, but I don't think she actually noticed any shockingly unusual talent in me. Every kid has a little bit of a mathematician in them. All kids know that some things are big and some are small. All kids know the concept of *more*. And kids learn early that logic—reasoning—has power, even over their parents. They find flaws and loopholes in their parents' rules. They can see that some things are necessary but not sufficient, that a particular statement was not applicable to every situation, that certain assumptions imply others, even if they don't know what any of those words mean. They begin to develop a grasp of the abstract. They ask why, and they want answers. I wasn't really different, just probably more willful.

AFTER I STARTED at the local public elementary school, my mom and I would race each other in sudoku at the kitchen table. One Saturday she took me to Toys "R" Us—it was the mid-1990s—and filled the cart with a couple of plastic action figures and a stack of boxes: Battleship, Connect Four, Othello, Monopoly, Sorry. We spent the rest of the afternoon playing. After that, every Friday night was Family Night, when we would order pizza and play board games at the kitchen table or on the living room floor. It didn't matter to me that my mother was older or had more experience; as soon as I understood the rules, I expected myself to win. Losing left me angry and frustrated to the point of pain. So I learned to become a little exploitative landlord in Monopoly, building up property as fast as I could, trying to bankrupt her. I never let her fold; I wanted to play out every victory. When we played Connect Four, I would stare at the empty holes in the plastic frame and see ghostly patterns of black and red. After I turned eight, it was another twenty years before I lost a game of Connect Four again. We played chess too, at a little chess table in the corner, and I started to see how combat could be a kind of dance, each move inviting and resisting the next. If my mother was proud of me for beating her, she did a good job of hiding it. She was nearly as competitive as I was. Still, no matter how many times I won, she kept playing with me and encouraging me.

My mother, of course, was doing all this to give me a jump on school, but I had no idea of her intentions. I hated school. In the classroom, I was bored and sullen. My mind wandered. I spent half the day counting the flecks in the linoleum floor. Sometimes I wasn't aware that I was being called upon.

One day my first grade teacher called my mother into school to tell her that I was having problems "processing," and that they should consider holding me back a year.

My mother was shocked. *That boy is finishing workbooks at home for third and fourth graders*, she told my teacher.

My teacher doubted it.

My mother looked at my teacher and got an uneasy feeling about what was going on. She knew that I had been finishing math books faster than she could buy them. But she could tell from my teacher's words and tone that my teacher saw me as a child from a single-parent home and viewed my shyness as a sign that I was a typical minority student unable to keep up in a classroom setting.

Test him, my mother said.

And so my teacher did. I had no idea what was going on. My mother didn't tell me that my teacher was concerned with my performance. She also didn't tell me that I'd done so well on the test that the teacher wanted me to skip a grade instead of making me repeat one. *No thanks*, my mother told her. I was young for my year as it was, and she wanted me to be among children my own age. After that, she was convinced that I needed to go to a private school, the kind of place that touted its diversity and progressive pedagogy. She wanted me to be in a place that would not be biased against me. She didn't want me labeled as a "lesser than," or to feel that I could not follow any path I wanted because of my background.

She succeeded. I had no awareness of prejudice, no sense that I might be limited by the color of my skin or living with one parent. I just kept happily doing the math work and puzzle books. While she prepared briefs and pored over depositions in the guest bedroom that she used as her home office, preparing to try a case, I sat in a black swivel office chair at a desk in my bedroom, my feet dangling in the air. I could do math puzzles for hours without noticing the time slip by. To me, it was the same as playing.

My mother slyly encouraged me in other ways. Instead of giving me an allowance for making my bed or taking out the trash, when we went to the store, she would let me keep the change if I could calculate her change before the cashier gave it to her. Pretty quickly, I became quite good at that. To protect her pocketbook, she upped the challenge. Instead of calculating the change in my head, she had me calculate the sales tax before the cashier rang up the items. The sales tax in Buffalo at the time was 8 percent, so I had to add up the costs of all the items and then calculate 8 percent of the total before the cashier was finished ringing up the items. I became very, very adept at multiplication that involved the number eight.

Because I had such a short time to do the mental math, I had to come up with an easier and quicker way to do it than adding or multiplying or dividing long strings of numbers in my head. Instinctively, I started looking for shortcuts or tricks. I figured out that if I broke numbers apart or turned them into numbers that I could calculate automatically, then I could arrive at the right answer with a lot less trouble. For instance, I didn't know what 19×8 was off the top of my head, but I figured out a way to do it just as fast: nineteen eights is the same as twenty eights minus one of them. Since I knew that $20 \times 8 = 160$, all I had to do was subtract 8 to get the answer. Doing that isn't hard at all—even if it looked that way. Cashiers would stare at me in amazement. My mom knew better; she knew my tricks. Before too long, she announced that we were done with that game. I was draining her bank account too fast.

When my mother argued cases in night court, she would bring me with her and seat me in the back of the courtroom with a copy of *The Big Book of Tell Me Why*, by Arkady Leokum. It was filled with the kinds of questions that kids ask and parents can't answer. *How do spiders spin their webs? Why does the moon shine? How does lightning work?* I read it over and over. That might seem odd—it's not like the explanations in it ever changed. But it never bored me. The fact that the answers to those questions were always the same was part of their appeal. They were facts, solid and demonstrable. They gave me a sense of power. They helped me understand why the world worked the way it did.

I liked that clarity, even craved it, because sometimes the rest of life was much too confusing. My teachers at my new school viewed me with less skepticism than those at the old one had, but I was not exactly an inspired student. I had no idea how to talk to them—or, more important, to my classmates. No one ever picked me as a partner. I had no friends. One day at recess, two of the kids who liked to make fun of me most backed me into a tree. *Lick it,* they ordered.

No, I said lamely.

Lick it!

It didn't seem to matter that I was bigger than they were. While all the other kids were watching, I turned around and quickly ran my tongue over the rough, horrible bark.

But at home, doing workbooks, I could forget all about that. I liked problems involving special or surprising properties of mathematics, puzzles that used concepts such as prime numbers or perfect squares, which are integers (whole numbers) that can be written as a number squared (4, for instance, is a perfect square, because $2^2 = 4$). I was intrigued by magic squares, grids of numbers in which no number is repeated and all the numbers in every direction (horizontally, vertically, and diagonally) add up to the same number.

$$\begin{array}{ccc} 4 & 9 & 2 \\ 3 & 5 & 7 \\ 8 & 1 & 6 \end{array}$$

I especially loved one type of problem: logic puzzles. To me, they were games, detective stories, treasure hunts. Given a bunch of clues, I would have to deduce the conclusion. I liked knowing there *was* a conclusion. There was clarity. On any given day, I ran into a lot of problems that I couldn't solve, like what to say to my classmates on the playground. Puzzles were simple and rule based. There were answers, and I could find them. They only seemed messy, a tangle of information shot through with gaps. But I could untie the knots and use the thread to connect the information. I never read the lessons at the start of each section or the instructions for completing problems. If I didn't figure out how to approach it myself, I thought, then I wasn't really solving the problem. I didn't like being helped. If I needed a map to guide me, then I wanted to be the one to draw it. Puzzles gave me a glimpse of a world, unlike the one I lived in every day, that I could make sense of.

ONE AFTERNOON, while going through a puzzle book and listening to Fiona Apple—my favorite musician—I came across a puzzle called the Einstein puzzle. The introduction mentioned a legend that Albert Einstein used it to test prospective graduate students who wanted to work with him. I've since read that the puzzle is named after Einstein because he invented it as a boy. Other versions call it the zebra puzzle, and say that it was invented by Charles Lutwidge Dodgson, a mathematician at Oxford better known as Lewis Carroll, author of *Alice's Adventures in Wonderland*. I've also seen it said that 2 percent of the population can solve it, a fact that seems to have as much truth behind it as the Cheshire Cat. The different versions use different clues for the problem, but the basic scheme is always the one that I encountered as a kid.

The puzzle goes like this: five people, each of a different nationality, live in five adjacent houses, and each house is a different color. Each person owns a different pet, drinks a different beverage, and smokes a different type of cigarette. You have fifteen pieces of information about the group.

1. The Englishman lives in the red house.
2. The Swede keeps dogs.
3. The Dane drinks tea.
4. The green house is just to the left of the white one.
5. The owner of the green house drinks coffee.
6. The Pall Mall smoker keeps birds.
7. The owner of the yellow house smokes Dunhills.
8. The man in the center house drinks milk.
9. The Norwegian lives in the first house.
10. The blend smoker has a neighbor who keeps cats.
11. The man who smokes Blue Masters drinks beer.
12. The man who keeps horses lives next to the Dunhill smoker.
13. The German smokes Prince cigarettes.
14. The Norwegian lives next to the blue house.
15. The blend smoker has a neighbor who drinks water.

The challenge is to deduce who owns the fish.

At first, all these facts seemed random and disconnected. If I had only glanced at the list, I probably would have thought there was no way to make sense of the information. But as I studied it, I began to see a beautiful structure, a way to join the disparate facts so that the full picture would emerge. It was like a jigsaw puzzle where some pieces were missing—but it was possible to re-create them by looking at the surrounding landscape. So I visualized the problem. I imagined it as a row of houses.

In my head, I also created a grid. There were slots for each piece of information, beginning with the given facts: the five colors of the houses (red, green, white, yellow, blue); the five different pets (dogs, birds, cats, fish, horses); the five types of drinks (coffee, tea, beer, milk, water); and the five brands of cigarettes (Pall Mall, Dunhill, blend, Blue Masters, and Prince).

Then, mentally, I started to organize the clues in relation to one another, to see what was hidden between the lines. When I started doing these kinds of logic puzzles, I used a pen and paper to make the table, but before long I was able to do it entirely in my mind. I wanted to carry everything I needed to solve a problem in my head.

I knew that the Norwegian lives in the first house (clue 9), and I also knew that the Norwegian lives next to the blue house (clue 14)—which means the second house is blue (since there is only one house next to the first house). Clue 8 tells us that the man in the third house drinks milk. So from those clues alone, I knew something about the first three houses. But that's not all that the clues told me about those houses. Each clue had in it a half-hidden door into another. All of the information I needed was there, and it was up to me to find it. For instance, since the green and white houses are next to each other, and the second house is blue, the first house can't be white or green. This implies another fact:

		COLOR					NATIONALITY					PETS					DRINKS					SMOKES				
		yellow	blue	red	green	white	Norwegian	Dane	English	German	Swede	cats	horses	birds	fish	dogs	water	tea	milk	coffee	beer	Dunhills	blend	Pall Malls	Prince	Blue Masters
HOUSE	first																									
	second																									
	third																									
	fourth																									
	fifth																									
NATIONALITY	Norwegian																									
	Dane																									
	English																									
	German																									
	Swede																									
PETS	cats																									
	horses																									
	birds																									
	fish																									
	dogs																									
DRINKS	water																									
	tea																									
	milk																									
	coffee																									
	beer																									
SMOKES	Dunhills																									
	blend																									
	Pall Malls																									
	Prince																									
	Blue Masters																									

the first house is either yellow or red. That means that the Norwegian lives in a yellow or red house. But I already knew that the red house is occupied by a Brit, which means that the Norwegian lives in the yellow house—and so he smokes Dunhills (clue 7), and he lives next to the man who keeps horses (clue 12). And so on.

Methodically, I filled in the chart in my head. I felt like Sherlock Holmes, hunting for evidence and making deductions, seeing facts that were invisible to ordinary eyes. I worked through the clues with an increasing sense of excitement and a growing sense of mastery. With a burst of triumphant pride, I concluded that the German owned the fish.

Why, you might wonder, would I want to spend even a minute doing this? Do Blue Masters cigarettes even exist anymore? What does this kind of puzzle have to do with *anything*?

When I was a kid, I would have just shrugged at the question. Why did puzzles need to be useful? To me, they were a game. Games didn't need to be justified. I didn't worry about whether the puzzles would be useful; I just thought they were fun. Now I hear the question a little differently. It's a variation of the complaint that math teachers hear from their students constantly. Why waste time figuring out where a fish lives? Why spend hours doing homework calculating differential equations? When are most people ever going to use that in life?

The honest answer is *never*. Most adults don't need to know calculus. They don't need to know how to calculate the roots of a quadratic equation. In all but a few lines of work, they can get by very well without even knowing what a quadratic equation is.

But it would be a mistake to think that they don't use math, or that figuring out where a fish lives has nothing to do with real life. When I was doing that logic puzzle, I was training my mind. I was learning how to separate variables, recognize patterns, identify relevant information, and create my own tools and techniques for solving problems. I was beginning to understand when to trust my instincts and when to double back. No matter what career you have, you need to be able to reason well, recognize patterns, use logic, compute numbers, make rational predictions, and communicate your thinking clearly. Everybody needs to solve problems.

SOME OF THE WAYS that math is necessary are pretty obvious. When my mom was going around the store with me as a nine- or ten-year-old, she knew calculating the tax of the bill wasn't going to be a necessary skill for me to learn. But she did know that I would need a sense of how things added up before she got to the cash register. She also needed to keep in mind what exactly the price represented. *Buy in bulk*, my mom instructed me early on, pointing out that one box of cereal might look cheaper than another, but what really counted was the price by the ounce. Or, at the drug store, she'd show me that different medicines—one a name brand and one a generic—were actually the same product, with the same active ingredient, only with different labels and very different price tags. It might not sound like that involves math, but it requires a mathematical concept: if two things with different names are equivalent to the same thing, then they are equivalent to each other. Those are simple examples, but a grasp of the properties of logic and sound reasoning has helped me make countless decisions, big and small, throughout my life. It has helped me weigh costs and benefits, distinguish between causality and correlation, move between the abstract and the concrete, and make hard choices. Thinking like a mathematician doesn't necessarily mean tackling a math problem.

That's not why I love math, though. I am still drawn to it for some of the same reasons that appealed to me as a kid—back when I was doing the Einstein puzzle and playing with magic squares, long before I had heard of a partial differential equation. All those puzzles were giving me a glimpse of the possible pathways of thought. They were revealing the threads that bind together music, language, maps, mechanics, sports. The shapes of things, the flow of liquids, the orbits of planets, the arc of a thrown ball. The transmission of information and the strategies of war. I could look around my house and see not a brick box, decorated in shades of brown and amber, but a series of solved math problems. Each one had involved a challenge.

I wanted challenges. I liked the feeling of being tested—even if I disliked the tests we took at school. Improvement did not always come easily. It took work. But there was nothing like realizing that what had seemed hard before now seemed easier, or that what I had done badly before I could now do well. Solving problems like the Einstein puzzle gave me satisfaction and clarity I rarely felt anywhere else. It gave me a sense of purpose. It gave me a sense of power.

Two

The Photograph

Football, 1996–2005

My first memory of football is from when I was about five years old. It is not of playing or watching the game. It is a photograph of my father, propped on a shelf in his home office. I would stare at it when I visited him at his house in Canada, an hour over the border from Buffalo. It seemed barely possible to connect the slender, serious surgeon I knew, the distant man who was working at the desk behind me, with the solid young man in the faded picture—his broad shoulders exaggerated by the pads, his uniform dirty, his expression defiant.

My father had played linebacker for the University of Alberta, not too far from where he grew up. He wasn't a sentimental person and didn't talk about his playing days much. Football to him wasn't something that made his eyes mist over. He didn't consider it noble. He knew how brutal the sport was. My dad never romanticized football for me. He knew it was a rough, violent game. *Coaches don't make men*, he sometimes said. *That's a load of baloney. They make football players.* Still, he loved football.

I loved it too, though not because it was something I experienced with my father. We didn't watch much of it together. On Saturday mornings, he would drive across the border, pick me up, and bring me back to his house in Grimsby, Ontario, or to his office at the hospital, where he was a thoracic surgeon. I would entertain myself while he worked, and then he would take me to the mall, where there was a Borders bookstore where he would read while I wandered, picking up whatever caught my eye. We would finish the day at the french fry stand.

When I was in the fifth grade, in 2002, my father moved to Boston to take a job as the head of thoracic surgery at a top hospital, Beth Israel Deaconess. He drove back every second Saturday to visit me, but I spent more time thinking about him than seeing him. I wanted to be like him. I thought of that old photo, which made me want to play football.

But I also wanted to play football because, once I got to middle school, I wanted to be like other kids. I was very purposeful about it. I'm not too proud of this. There are some children who are a little different and who can do their thing and ignore the taunts, who don't mind too much being the last one chosen for teams, the one no one wants to partner with. I admire those kids—but I was not one of them. I wanted friends, and when I reached middle school, I wanted to fit in. So I looked at what the other kids said and did, and I tried to copy them. Conformity was a beautiful thing.

The other kids played street hockey and video games, so I played street hockey and video games. They played soccer and lacrosse, so I played soccer and lacrosse. They watched the Buffalo Bills game every Sunday during the fall. So did I. They played Pop Warner football, and I wanted to play too. But Pop Warner had a weight limit, and I was too heavy. When I tried to join the middle school's team, the school couldn't find a helmet that would fit, because my head was too big. In hindsight, it was for the best. Not getting to play football as a kid did not hold me back later on, and I was probably lucky not to play tackle football before high school, while my body and brain were still developing.

Plus, I wasn't a very talented athlete as a kid. For starters, I was overweight. I was big in every way—by the time I was in eighth grade, I was six feet tall—but I was also out of shape, too weak for my size. The only thing that set me apart from other kids when I played sports was my intensity as a competitor. I couldn't stand losing—so much so that I would do everything in my power to try to win. I was willing to exhaust myself and to take risks that other kids wouldn't. I took things seriously even when no one else did.

My soccer team was in one of those leagues that only gives participation trophies. The coaches didn't even keep track of the score during games—but *I* did. If my team lost, I'd come home in a furious mood. My mother would try to remind me that the game was just for fun, that there weren't winners and losers. *That's stupid*, I'd grumble. *Someone always wins!*

When I was in seventh grade, in 2004, my father moved back to the Buffalo area. Burned out from the intense, unforgiving life of a thoracic surgeon, he decided to leave medicine and pursue his other interests. He enrolled in a master's program in economics at the University of Buffalo, which also allowed him to spend more time with me. When he got to town, he took one look at me and narrowed his eyes. I could see him mentally weighing me. *You're carrying too much weight,* he told me. I didn't say anything. I was already a little self-conscious about my size.

My dad decided that he was going to do something about it. When there was an elevator, he would say, *We're taking the stairs.* He would pick me up from school and take me first to the university library, where both of us would do homework, and then to the university gym. He put me on a program, lifting weights and running sprints or stairs. I complained, a lot. My dad wasn't the kind of guy who would let up because I was whining. He was in great shape, and at the gym he showed it. *I'm never going to be as strong as you,* I said one afternoon.

He looked at me—my big bones, my broad frame—and paused. My mother was nearly six feet tall and broad-shouldered and had the loose grace of a basketball player. At thirteen, I was already almost as tall as he was. *Eight months,* he replied. *You'll be lifting more than me in about eight months.*

Three

||||||||||||||||||||||||||||||||||||||||||||||||||||||||||

Crashing Calculus Class

Math, 2004–2009

When my dad and I didn't go to the gym, we'd go to the library at the University of Buffalo before I headed home to my mom's. One day when I was in eighth grade, while I was doing some homework and my dad was working on a problem set for his economics master's program, I asked him what he was working on. *This is a matrix,* he said, showing me a rectangular arrangement of numbers in rows and columns held together by long brackets on the left and right. Then he pointed to the numbers inside the brackets. *The numbers are the elements of the matrix.* He explained how to add, subtract, and multiply different matrices together. *They're pretty useful,* he said. *See how this matrix is shaped like a square? We can learn some of its properties by calculating something called the determinant. That's what I'm working on now.*

How do you calculate the determinant? I asked. He showed me how to use the well-defined determinant of a 2 × 2 matrix—a matrix with two rows and two columns—to determine the determinant of a matrix of any size.

A few minutes later, I was playing with matrices—adding them together, turning them upside down, and trying my dad's homework problems myself. He stared at me, surprised.

That summer before eighth grade, my mother enrolled me in a summer camp for kids who were interested in engineering, as she had for several summers in a row. This year, though, I was bored. I liked the other kids in the class, but I couldn't understand their enthusiasm for the projects we were doing. We were making rockets propelled by vinegar and baking soda, and building models with balsa wood and glue. Everyone seemed to be having fun except me. I was miserable. It seemed to me that we weren't doing *real* math or science. We weren't learning anything exciting or difficult or new. We certainly weren't being challenged in the way that I expected and wanted to be challenged. Baking-soda rockets? We might as well have been following the instructions from a kit available at a toy store. I had no problem with games—I loved games—but the projects we were given made me feel as if we were being treated like children.

I have an idea, my dad said after I complained to him. He used his University of Buffalo student ID to enroll me as an auditor in a calculus course for business students during the summer semester. His name was John Urschel; my name was John Urschel. And I could pass for someone much older than my age, since I was nearly six feet tall.

There were about thirty people in the class, mostly business majors. The course was taught by a graduate student who seemed almost as nervous as I was. He would try, awkwardly, to connect with the students by talking about muscle cars before the start of class. The other students mostly ignored him—and, for a while, ignored me. My nerves at being found out quickly disappeared. I thrilled to the challenge of fitting in—to doing the work well enough that no one was the wiser. I liked the anonymity of passing for someone older, someone else— another John Urschel, a John Urschel who could have been any regular college student.

Except that your average college student probably does not enjoy doing calculus homework.

Learning calculus was like learning a secret code. If you were ignorant of it, much of the physical world was indecipherable. But if you knew calculus, you could describe the orbit of planets or the spiral of a football. It didn't occur to me that calculus was supposed to be too hard for me—and it wasn't. I'm convinced that most other thirteen-year-olds wouldn't find it too hard either, if they weren't conditioned to think that calculus was way too advanced for them. In fact, a ten-year-old could understand the underlying themes, if not perform the calculations. The question that calculus begins with is basic: what happens if we think of a smooth curve as a straight line?

Just that question was enough to make some of the mathphobic people in my class turn off their brains. You could see their eyes glaze over, even before they had to solve a single problem. It sounds like a contradiction: a curve as a straight line? But it's very easy to visualize. I imagined a cannonball shot straight across the face of the earth. I would be able to see only a short bit of its path through the air as it flew by, and so to my eye it would look like it was following a flat trajectory. If it were possible to see the cannonball from a long distance away, though, I would be able to see its whole path. As long as it didn't rip into something, eventually I'd be able to see the projectile curve toward the ground as gravity pulled it toward the earth (which, of course, itself looks flat but is actually round). Then I thought of a football moving through the air. I know that a thrown football follows an arc. What if I could see only a small segment of its path? If it was thrown hard and directly, it would look like it was moving in a straight line. If you drew the arc of a football on a piece of paper and then zoomed in close to a segment of the arc, the same thing would happen: the curve would flatten, and then flatten more, until it looked like a straight line. The closer you zoomed, the straighter the line would be. If you zoomed in so close that the segment was infinitesimally small— smaller than any size you could imagine, but bigger than zero—then it turns out the curve wouldn't just look kind of like a straight line. It would be infinitesimally close to one.

That, more or less, is what Isaac Newton realized, back in the seventeenth century. The slope of that line (a calculation of its steepness) is called the derivative, and it forms the basis of one of the two main branches of calculus, differential calculus. (The process of finding the derivative is called differentiation.)

I learned that the derivative can be a very useful way of measuring change. It lets us relate the position, velocity, and acceleration of a football—or an airplane, a missile, a planet, a speck of dust. Or— since I was in an economics class—it lets us represent and explain economic behavior, like cost minimization and profit maximization. The other branch of calculus, integral calculus, is related to differential calculus (that, in fact, is the fundamental theorem of calculus), but it deals with questions of area, volume, and displacement, by letting us calculate the area under a curve. It's fairly easy to calculate the area of anything bounded by straight lines—just divide it into a bunch of triangles—but harder when you're dealing with the area under, say, a roller coaster. It turns out that you can divide the area under curves into a bunch of infinitesimally thin rectangles and then add them all together. Calculus, I quickly grasped, let me move from a world that was static and frozen to a world that could move and flow. It was as if I had stepped out of a photograph and into a movie. Suddenly, time could pass, speed could quicken, gravity could pull, economies could grow.

It was easier to understand calculus as another language because some of it *was* in another language. There were new symbols to learn, some of them from Greek. The countless hours of doing puzzles and brain teasers had helped me learn to move between numeric and symbolic thinking. Besides, there was something exciting to a kid about seeing math as a kind of code. Learning it would give me the ability to describe a half-hidden world—*our* world, which we rarely stop to see. It wasn't that it was extremely easy for me to compute differential equations and integrals. I struggled with the more advanced techniques as did everyone else. But I hadn't internalized the common idea that the word "calculus" is basically a synonym for something impossible. Instead, I saw it as a way of describing the physical world as we all experience it. Calculus was a way of turning the physical world into puzzles, and I had never stopped loving puzzles.

I couldn't solve the puzzles fully yet, of course. I didn't understand everything in that class. Totally grasping the idea of something infinitesimally small or thin, for instance, or adding infinitely many things together, required an understanding of the concept of infinity that I would not achieve until college—if then. I took for granted that I knew what "infinity" was, until I was forced to manipulate it in more advanced ways. But I didn't need that level of comprehension to do well in calculus. I thought the subject was accessible, in part because I was too young to have it hammered into me that it shouldn't be. I didn't dread calculus. It excited me. It felt like *real* math. It was revealing something fundamental about the world. And perhaps it revealed something fundamental about me too. I wanted that kind of challenge. I liked countering expectations. I wanted to do things that people assumed someone like me—in this case, a kid who hadn't even started high school—couldn't do. And I wanted to learn things that were rigorous and interesting and surprising. I was willing to put in the work.

I would sit quietly in the back of the classroom, hoping not to attract attention. I didn't try to make friends. But after a while, a few of the other students started to notice that I was doing well. When the teacher called on me, I was ready with an answer. When he passed back problem sets and exams, my grades were consistently high. One day, as I was waiting for a lecture to start, one of my classmates came over and spoke to me. *Hey man*, he said, *did you finish the problem set yet? I could use some help.* He slid into the chair next to me, and I walked him through the work. As the semester wore on, more students came over to me and asked me for help with homework or to clarify something. No one seemed to notice anything unusual about me, except that I had a good grasp of the coursework—which is how I liked it.

||||||||||||||||||||||||||||||||||||||||||||||||||||||

Four

Canisius

Football, 2005–2008

That fall, I entered Canisius, an all-boys Jesuit high school. The school occupied an old, ornate mansion in a historic district of Buffalo. It had gables and turret-like chimneys—an early-twentieth-century bank president's idea of a castle. When the Masons bought it in the mid-1900s, they installed a bowling alley and Turkish baths. But when the Jesuits took over and turned it into a school, it became a place where shirts with collars and ties were required, a place where discipline was strictly enforced.

Canisius was an academic powerhouse, and I was there on an academic scholarship, which I had been awarded based on the results of the entrance exam. But I barely cared about schoolwork. The thrill I had felt during the calculus class at the University of Buffalo faded almost immediately. The John Urschel who was proud of passing as a regular college student was replaced by the John Urschel who just wanted to be a regular high school one. I did what was asked of me, coasting through my math and science classes and getting by in the rest, with some help from SparkNotes. The only thing I really cared about was football.

When I went out for the team as a freshman, I was one of the youngest, and one of the few with no formal football training experience. Most of the other freshmen had played Pop Warner. On the first day of training camp in August, the equipment manager dumped a pile of gear into my arms. I went back to my locker and tried to untangle the straps and laces. It was getting late; other players were already heading out to the field, their cleats clicking the floor like a timer set to warn me. I tried watching the other guys to see how they were doing it, but they tightened their straps and laces as easily and naturally as if they were tying a shoe. Somehow, I managed to struggle into my pads and make it onto the field, though everything felt tight and loose in the wrong places. I was sweating before I ran a single step.

If my coaches noticed how awkwardly I was moving, they didn't let on. I felt their gazes lingering on me as they looked around the group of freshmen. It was easy to guess why: I was bigger than pretty much everyone else. When the teams were announced, I was designated as a lineman and put on junior varsity instead of the freshman team.

When we started drills, a coach pulled me aside and taught me how to hit. He set my feet at shoulder width and showed me how to balance, so that my weight was loaded but light on the balls of my feet, my back firm, my eyes forward, my hands ready to grab and push. Then he showed me how to explode out of my stance, driving my arms and knees high and forward, staying low to use my leverage and unbalance my opponent, accelerating through the tackling dummy.

Don't slow down, he barked. *Don't be soft.* He showed me how to wrap my arms around the dummy and rotate and wrestle it toward the ground. And once I'd mastered that, he put me against a lineman. *Explode through him*, he instructed.

I went down into my stance and mentally checked the distance of my feet, the angle of my back, the position of my hand, running through the instructions in my head. *Don't grab his jersey. Don't slow on contact. Come on, John*, I whispered to myself. Everything seemed to fade away—the sound of the other players, the heat of the high sun, the sight of the school beyond the field. And then the contact came, and the grunts as the air left our bodies, and I could hear the crash of the pads against pads. I could feel my feet slipping beneath me and willed my legs to hold the force as I tried to twist him to the ground. The smack was a shock, the blow unlike anything I'd ever felt. It was exhilarating. I felt a rush of aggression. It was like the testosterone-fueled teenage anger that, in daily life, I managed to hide. Only now, within the clean white lines marking the field, it was not only allowed, it was unleashed.

As a sophomore, I made the varsity team. On the first play of the first game of the season, the coaches tapped me to head out with the kickoff coverage team. As the ball soared through the air, turning end over end, I sprinted the length of the field. The returner stuttered and spun, but I launched myself into him and made the tackle.

On the next kickoff, I did it again. My dad took note—literally. After that, he would sit high in the stands with a pair of binoculars and a pad of paper and a pen. After every game, we would go out to dinner and he would run through his notes, telling me where he thought I'd done well and where I'd messed up, what I needed to work on and how. Often, he would show up to watch practice.

By my junior year, I was playing offense and defense—offensive and defensive lineman—and special teams. When I got tired, I'd tap my helmet as a signal to the coaches to take me out. They usually pretended not to see me.

By the end of my sophomore season, my dad was convinced that I had the potential to be a good football player—maybe a very good football player. He knew my natural strength and saw that my body was still filling out. He noticed my balance and footwork, and knew from watching me play lacrosse that I could move and block people out. (The skills of a lacrosse defender translate to pass blocking fairly well.) He started telling me about the nuances of different football schemes. At home, we'd watch tapes of old Super Bowls and would study the old-school techniques. Some of the most helpful advice he gave had more to do with the psychology of the game than with any technical skills. *Once during the first quarter, on a run play, ignore the run and pancake a guy in the backfield,* he would tell me. *Do it where his teammates can see it. The other guys will be looking at him, wondering what happened, and he'll be stunned. Believe me, he won't be the same player for the rest of the game.*

That summer, he bought me pads so that I could train in the backyard of his house—and he bought a set for himself too. He would show me different blocking techniques, including a few that have fallen out of favor. One day, he put me on my back a few times. I stared at him, half-dazed in amazement. By then, I was bigger and stronger than him; he wasn't supposed to be able to beat me. *How did you do that?* I asked.

So he showed me. He had taken the crown of his head and smashed it into mine. These days, because of concussion concerns, the head is supposed to remain as untouched as possible. But he knew what worked: take the defender head-on. He was basically using his skull as a battering ram. *You can teach the other guy a lesson*, he said. *Pop him back.* That was the way he'd learned to play, and that's the way he taught me to play.

He taught me other things—to be nice to the refs, for instance. It could make a difference on a close call if they thought you were a nice guy. He taught me to help a guy up after I'd put him on his back, because it was the right thing to do. He taught me what he knew.

FOOTBALL GAVE ME A WAY of being in the world. It instilled in me a kind of confidence that I had never felt before, a confidence that I could carry off the field. It helped me feel like I fit in. I made friends easily at Canisius. We would stay up all night playing video games and watching movies, talking about girls from other schools, or, later, raiding their parents' liquor cabinets. We supported one another and annoyed one another and had a lot of (probably too much) fun. Football showed me how to be a teammate, which helped me be a friend. It also taught me toughness. I practiced with an intensity that few other guys matched. That was the nature of the game, the risk we took. I knew that if I let up, the person on the stretcher could be me.

In the rest of my daily life, I had to restrain the competitiveness I'd felt ever since I'd played Battleship with my mother as a child. Football broke the dam. During preseason practice, the coaches would sometimes pair us off and tell us to kneel, facing each other. We'd wait, soaked with sweat after a hard practice, staring at each other—teammates at any other moment, but opponents now. A coach would walk by and hand each pair a towel. Both players would hold it, one hand on the inside, one on the outside. The goal was to win the towel from the other person. I would dive, twist, pull, wrench, tear—every cell of me straining. I would have done almost anything to come away with the towel.

In a weird way, those combative moments made me and my teammates closer. We fought one another in practice so that we could help one another in games. But in high school football, what I really loved was blocking. I had the right body to be a lineman, and I had the right temperament. I was generally seen as a nice person—I won a lot of sportsmanship awards, mostly because I would pull my opponent up off the ground, the way my father taught me—but I didn't feel nice. I was a teenage boy, and I had a lot of aggression. Mostly, I managed to hide it, but it was always there. Football helped me channel that feeling and release it. I would feel a surge of power as I slammed into the body of another player, and a sense of dominance when I knocked him over. I loved the sound of pads crashing together. I took it for granted that there was something good about the sound of a grunt as breath left the body—whether from effort or because it had been crushed out. I believed that toughness was a virtue. I wanted to be tough, in some simple and obvious way, and to feel the rawness of what I felt only on a football field.

I didn't know anything else like that rush. It felt almost forbidden—and in some ways it was. On a football field, you could do things that you couldn't do anywhere else in life. The truth is that few other feelings were so good—at least to a teenage boy like me. Male aggression can lead to huge problems in society, obviously. The whole point of civilization is to restrain and erase those instincts that drove me to crash into and overpower people every Saturday afternoon. But the white lines demarcating the football field actually gave me a sense of limits instead of blurring them. There was a place where aggression and the surge of desire for dominance were allowed, a place where those impulses could even be productive. We put our animal instincts toward some common human purpose. When I stepped off the field, I left the place where that was permissible, but I knew I would be coming back. For some guys, maybe it's different. Maybe the distinction between football and the rest of life becomes hard to see, and they can't turn their aggressive instincts off. But for me, in high school it was easier to be calm off the field if I could pretend to be a gladiator on it.

In the off-season, I had planned on playing lacrosse, but one day the track coach, Brian Lombardo, spotted me in the hall and got the field coach to persuade me to throw discus and shotput. Lombardo was an elite runner—young, long, and skinny, with pale skin and close-cropped hair. He became more than a coach to me, and even more a mentor than a friend. He was someone I could talk to about anything. But football was what captured my imagination.

My dad was right when he said there was nothing noble about football. Still, there were things I learned, lessons I kept with me. I embraced the little rituals of being an athlete, studying plays, solemnly wrapping my wrists before games. Every Friday night during the fall, I watched the movie *Friday Night Lights*. I loved that movie. I recognized in it the grit of high school football—the physical pain, the dedication, the rivalries on the field and inside our own locker room. I understood the truth of its message, however sentimental it sounded. I knew how much winning mattered—more than anything—and yet I knew how fleeting the feeling was that winning brought on. I hated losing and *needed* to win, but the elation or disappointment it produced would fade. It would turn out that everything, including myself, was just the same.

|||||||||||||||||||||||||||||||||||||||||||||||||||||||||||||

Five

Rocket Science

Math, 2007–2008

During my sophomore year, I was offered the chance to do a more accelerated math program at the University of Buffalo. I turned it down. It would have interfered with football practice. I treated schoolwork, for the most part, like a manageable distraction.

Sometimes, though, a subject would seize my interest. Something deep within me would stir, and I would feel that same sense of wonder and curiosity that I had felt doing puzzles as a child. *How do we measure time?* Mr. Magnuson, my physics teacher, asked during my junior year. It sounded like a question that I might have encountered in *The Big Book of Tell Me Why*.

Mr. Magnuson was teaching us the mathematics behind time dilation. The way he did it was not flashy. He did not show us catchy experiments or try to communicate just how weird the material was. He was not particularly enthusiastic or charismatic. He was moderately thin, moderately tall, and spoke in a monotone. The only thing that seemed particularly unusual about him was that he rode a bicycle to school. Some of my classmates fell asleep during his class.

I might have too. But for reasons I couldn't quite explain at the time, I heard something deep and resonant in his nondescript, matter-of-fact way of lecturing. I caught a glimpse of a world of forces, vibrations, and dynamics, all described in the language of math. It was rigorous and precise and unerring and everywhere—in the speed of a runner, in the fade of sunlight, in the passage of time.

What really caught my attention, though, was not that physics gave me a more precise way to describe what I already knew, but that it challenged me to rethink what I took for granted. When we talked about time, I had to question my assumption that time was immutable. I had thought a minute was a minute, measured in any way, and that a minute passed at the same rate everywhere. If two clocks disagreed, then it was the fault of their setting or faulty engineering, not the result of some deep underlying cause.

Most people, for most of history, assumed this. It turned out not to be true. Albert Einstein was concerned with how time looked to people, not how it drifted through the mind of God. From a young age, Einstein liked to come up with thought experiments that questioned untested assumptions or looked at the way the world works from a different angle. And what his thought experiments led him to suspect—and rigorous experimental results confirmed—was that time is relative. What matters is your frame of reference: time is moving differently for you if you're standing on the street than it is if you're in a moving car; it's different if you're in an office or on a spaceship. Neither time is "true," a reflection of "real" time. The effect is not the result of technical or mechanical aspects of the clocks, but of the nature of space-time itself. Time dilates. And even high school students can understand some of the math behind it.

One afternoon I spent some time on the internet reading more about how Einstein had thought through these problems—not just the content of his thoughts, but his approach. I was curious about the balance between calculation and conception, between reason and intuition. I found myself thinking more about time dilation as I walked between classes or waited for the bus. It was like a song stuck in my head. Einstein asked basic questions that no one else was asking. He looked for contradictions in traditional models. He used one idea as a step—or slingshot—into the next. His conclusions seemed totally crazy at the time. They challenged human intuition too drastically. A lot of people—including some of the smartest people of his generation— refused to accept them. And yet, his own intuition worked differently. He would reframe the problem, and viewed from the new perspective, his ideas made sense.

Physics was easily my favorite class. Part of that was probably because it rekindled the excitement that I had felt taking calculus as an eighth grader: here was a way to describe how the world actually worked. (Not coincidentally: classical physics and calculus share a father in Newton.) But part of it was because of the way Mr. Magnuson taught. He was knowledgeable and thoughtful, and—most important— he encouraged us to learn and think for ourselves. Most of my teachers would have said that was their goal, but very few of them actually gave us the opportunities and motivation to do it. Mr. Magnuson gave us opportunities to look beyond the material we were supposed to learn from lectures.

Once, Mr. Magnuson proposed an extra-credit problem that required some concepts that we hadn't covered in class. I spent weeks working on it obsessively. I would think that I had solved it and bring it in to Mr. Magnuson, and he'd gently tell me it wasn't quite right. I had to make my way to the solution methodically, certain of my understanding of each step. Finally, I came up with the correct answer, and the satisfaction I felt helped ensure that the love of solving problems I'd had as a kid did not die out while most of my attention was on other things. It also taught me the value of persistence, which has turned out to be one of my greatest strengths as a mathematician— my desire to prove myself against a problem, my determination to answer any challenge, and my resilience in the face of failure. It is a lot more valuable than being able to calculate large sums in my head.

From time to time, something like that extra-credit problem would grip me. For reasons that I could never quite explain, some concept would set my mind turning. I'd go to the library to find a more advanced trigonometry book than the one we used in class, and I would explore the properties of shapes. For a couple of weeks, I became fascinated by statistics, which also challenged my intuition. I was only sixteen, a junior in high school, but I was already becoming aware of how math like that could influence my reasoning.

For the most part, though, these bursts of curiosity were isolated. I kept my reading and thinking outside of class to myself. Not because it would have made me seem different, not anymore. My friends knew I was smart, and they were neither very impressed nor inclined to give me a very hard time about it. They didn't care. (Except when it came in handy. Sometimes they paid me ten bucks to finish their homework if they were in a rush.)

I didn't really care either. I knew that schoolwork came more easily to me than it did to the other guys, and I knew that I had an unusual aptitude for math and science. (History and English were another story.) I didn't make a big deal out of it, though, because it didn't seem to be a big deal. If I was curious about something, I looked it up. I always wanted to know the how and why of things for myself, instead of the rote learning we were expected to do in class. And I did well, because I was competitive—with myself and with others—so I always wanted to do the best that I could on tests. For the most part, though, I did my homework as fast as possible, without giving the material a second thought. And as my role on the football team grew, I cared less and less. As the time to apply for college approached, I didn't get more serious about academics. I wanted my football abilities to get me in.

My mother supported my playing football, as she always supported me. She helped organize spaghetti dinners on Friday nights before games. She set up bake sales and ran a concession stand at Canisius's football field to help raise money for the team. She would stand for much of the games, bouncing on the balls of her feet, cheering loudly in a low, full-throated voice. When a call went against me, she would scream at the refs, ranting that I had been wronged. I was her baby, and she was always ready to protect me. Besides, I knew that in her own way, she was as competitive as I was.

But in her mind, those Saturday high school games were as far as it went. Football was secondary to school. My ability and interest in math and physics, on the other hand, were not lost on her at all. During my junior year, she would spot me with a physics book in my hand, and we would sometimes talk about what I was learning in class at the dinner table or in the car. She had long been convinced of my talent. Now she could sense my interest. That was enough for her. She decided that I would become an engineer. Not just *any* kind of engineer. *I think you should be an aerospace engineer,* she told me one day.

You want me to be a rocket scientist? I replied, skeptical.

Yeah! A rocket scientist!

It was probably natural that my mother would want me to be a rocket scientist, and not just because it was synonymous with "genius" to her (which, since I was *her* child, she was convinced I was). Like many people, she assumed that the greatest questions and mysteries of the universe lay in space, and the greatest minds were those that dared to explore the vast dark unknown. It is a common fallacy to romanticize things that seem grand—and what could be grander than outer space?

Besides, she added, *there's always more to explore—which means good job security. NASA will always be hiring.*

It never occurred to her to encourage me to become a mathematician. Why would it? I doubt she had any better idea of a mathematician's career than she did of an offensive lineman's. I didn't really know what a mathematician did either. I never met anyone until college who would list it as a real profession. My middle school math teacher had a poster in her room of all the different careers that use mathematics: accountant, architect, banker, engineer, scientist, teacher—but no mathematician.

Of course, since I was her son, my mother thought I was supposed to be the greatest rocket scientist the world had ever seen. Naturally, I had to go to MIT. Princeton, or maybe Stanford, would be acceptable too.

But I had a different goal in mind. I didn't dream of being a rocket scientist. I wanted to be like Jake Long, the offensive tackle for the University of Michigan.

Six

‖‖‖

Recruitment

Football, 2008–2009

That desire, to be just like Jake Long, was absurd. At the end of my junior year of high school, in April 2008, Jake Long was picked first overall in the NFL draft. He was 6-foot-7 and 315 pounds. At the time, I was 6-foot-3 and about 220 pounds—nearly a hundred pounds less than Long and about eighty pounds less than the average big-time college offensive lineman. Recruiters prized size. They figured technique could be trained, but height was something that couldn't be taught. Offensive linemen needed to be over six feet and around three hundred pounds of mostly muscle. Size translated into reach, strength, leverage—basic physics. As defensive players got bigger and more athletic, offensive linemen needed to keep pace too. If a 6-foot-4, 310-pound defensive lineman was sprinting at you, and you were 220 pounds, it didn't matter how dedicated or disciplined or fearless you were. You were probably in a lot of trouble; he would probably run right through you. And that's before the quality of competition that I was facing in Western New York and, frankly, the quality of my skills were taken into account. My technique was awful.

Still, my football coach, Brandon Harris, told me I could play in the Big Ten. Coach Harris—short, African-American, with a thick neck, barrel chest, stubborn chin—was an enthusiast. *You have the work ethic,* he told me. *You have the talent!* I'd wake up in the morning and hear his words in my head. They'd pull me out of bed and into the weight room. Coach Harris was convinced—or at least he convinced me—that I had the potential to play for a big program. He put together a film of my highlights and sent it to coaches across the country, and made calls touting my accomplishments: captain of the football team, an all-state defensive tackle, and the Western New York lineman of the year.

If my teachers at Canisius could have heard him, they probably would have rolled their eyes. Football coaches have a tendency to take themselves and their mission very seriously, and their passion can come off as misguided or absurd, or even a little dangerous. Every year, there are probably thousands of high school players who assume they're going to get a football scholarship, and they blow off schoolwork and do stupid things because they think their future is set—and then they get injured or just passed over and end up working at a gas station. But the truth is, I wish some of my teachers had been more like my football coaches. I wish they'd shown the same passion about their subjects and had the same impulse to recognize and nurture potential. I excelled in math and science at an academic powerhouse, and my academic potential was clearly greater than my potential on the field, but none of my teachers ever told me that if I dedicated my time and my full focus to math or physics, then I could be a great mathematician or scientist. No one in high school ever called the

MIT or Princeton math departments and told them to recruit me. No one ever told me I could be Albert Einstein or John von Neumann, arguably the most brilliant mathematician of the twentieth century. (Nobody even told me who John von Neumann was.) I understand why they did not say those things: my teachers would have sounded ridiculous! But there is something to be said for having the imagination to aspire to the very highest goals, and for giving and getting the encouragement to commit oneself to get there.

DURING MY SENIOR YEAR, I started getting letters from football programs, big and small—including Stanford, which made my mom light up with hope. The phone began to ring. That didn't mean much, I knew. Schools send a lot of letters to a lot of players. The real interest in me wasn't coming from places like Michigan, where I dreamed of playing. Stanford stopped calling. The real interest—the kind of attention that would culminate in a scholarship offer—wasn't coming from anywhere except the University of Buffalo and the Ivy League. I was a two-star recruit, and top division-one teams, the kind of schools that get ranked and play in bowl games, were not exactly falling over themselves to offer me a scholarship.

My mother tried to take things into her own hands. She called MIT's football coach and asked him to recruit me. She even offered to send him my game film. The coach was a little confused. *That's not necessary,* he said. *If your son is admitted, he can play football. We don't cut anyone.*

Princeton, though, was one of the schools that recruited me for football, and so my mother held out hope that I would choose it. She so badly wanted the best for me. But I was resistant. However good the academics were there, it didn't feel like a place where it would be fun to play football, and that was what I really cared about. I visited and noticed the mild but pervasive sense of resentment among the guys. The team was insular—maybe understandably, because it was isolated.

My dad took me to the football stadium, where Princeton was taking on a winless Dartmouth squad in the last game of the season. The range of skills on the field was extreme. There were some excellent players—a few guys make it from the Ivy League to the NFL most years—who played with anticipation and awareness and a quickness that immediately set them apart. There were also players who looked like they could have played at Canisius, and a few of the backups I saw when they were put in at the end of the game might have even struggled to make Canisius's varsity. Guys missed tackles, dropped balls, failed to adjust to the other team's schemes. What struck me most, though, was not the quality of play but how empty the stadium was. The players were up for practice before the sun, just like every other player on every other team in the country, but when they played games, few students came to cheer them on.

You shouldn't go here, my dad said. *You won't be challenged to become a better player. You'll play in front of empty stands. This is what it will be like. No one will care.*

My dad wanted me to go to the University of Buffalo, not Princeton, since it played against much better competition than the Ivy League. He figured that I would get a decent education at any school—even if I had to give it to myself—and then I could go to a top school for a graduate degree. I was also considering Cornell, another Ivy League school, especially after I went on a recruiting visit. Unlike Princeton, none of the guys at Cornell seemed to have a chip on their shoulder. Cornell might not be a traditional football school, but its players know how to have a good time. I nearly committed on the spot.

How'd it go? My mom asked when I got back.

It was great, I answered with a smile.

Well, tell me about it!

Lots of Bible study, I said, and then went into my room and closed the door, ignoring her demands to tell her what I meant.

Then, one day in December, after the football season was over, Coach Harris poked his head into one of my classes. Everyone looked up; Coach did nothing subtly. *I need to see Urschel,* he said, ignoring the stares. *Now.* I slipped out of class, and as we walked toward his office, Harris told me that Mike McQueary, the wide receivers coach for Penn State, was there and wanted to see me. My tie started to feel tight around my neck.

McQueary was a former Penn State quarterback and he looked like it, even wearing a suit. I immediately noticed his broad shoulders, along with his electric orange hair. He was friendly and polite to me, but he did not overpower me with compliments. He had a kind of reserve about him. Penn State didn't exactly need me. It was clear that he held the cards. Most recruiting visits are chances for coaches to flatter and woo prospective players, but this one was different. The football season was long over, and Penn State had already made most of its offers to the incoming class of recruits. The top ones had signed the previous summer. McQueary wasn't there to sell the school to me. He just wanted to see me with his own eyes.

So I stood there, feeling sweat on my brow starting to gather, wishing I could loosen my collar, as McQueary looked me up and down, as he gauged the length of my arms and the size of my hands.

My mother was furious when I told her that my coach had me meet McQueary. *You should not be speaking to him,* she told me. *How dare they pull you out of class!* I could almost see the word *"Princeton"* printed across her worried forehead.

I could sympathize. She wanted the best for me. To her—a woman who had been the first in her family to attend college, who had to push for every opportunity she got—that meant, understandably, a Princeton education. But of course, I didn't listen to her. I was totally elated that Penn State was showing interest in me. I could not keep the smile off my face, which only made my mother more worried. But Mc-Queary's visit would mean nothing unless they offered me a place on the team.

LATE AT NIGHT a few weeks later, in mid-January, I was on my way home from a track-and-field meet when the phone rang. It was Mc-Queary. Penn State was offering me a full scholarship. I was the twenty-sixth of twenty-seven recruits for the incoming class.

I told McQueary that I wanted to accept the offer immediately, but I agreed to come visit first. A week later, my mom and I drove through the winding roads from Buffalo to State College. My trip was unlike any experience I'd ever had. Football schools have recruiting down to a science.

A place like Penn State doesn't just pitch players a program. It pitches a dream. Because the NCAA forbade schools from paying players, players saw none of the profits they helped generate in the form of wages. Schools can't compete for recruits by offering bigger salaries, so they compete by offering nicer locker rooms, fancier weight rooms, better sound systems, and meeting rooms with softer seats. They tailor each visit toward the recruit, to offer the best vision of a player's glorious future.

It's not all—or even mostly—cynical. Penn State players are more than happy to sell the school to recruits, because the vast majority are genuinely happy to be there. My host for the weekend was a great guy named Stefen Wisniewski, a talented offensive lineman whose father played in the NFL and whose uncle was one of the greatest players ever to come out of Penn State. Not incidentally, Wisniewski was also a three-time Academic All-American, an Honors College student with a 3.9 GPA. Not that any of that was apparent on my visit. We went to parties, we hung out with very friendly girls, and we did not go to class.

Penn State can promise something even most of the top programs can't. Football is like a religion in Happy Valley, and Beaver Stadium is its church. The stands hold more than a hundred thousand fans. On game days, the stadium becomes the fourth largest city in Pennsylvania. It was cold and empty when I walked around it. Even so, I could imagine the scene—the sound of the crowd, the sea of students in the stands, blue and white, and me on the grass. I had imagined it a thousand times, imagined standing at the center of something so much bigger than myself, and the sweat in my eyes and the pain, and the high-level, high-stakes competition. And here I was, almost there.

I'd already told McQueary that I was coming to Penn State, but at the brunch for the visiting recruits on Sunday of my official visit, he told me that I had to tell the coach, Joe Paterno, myself. Paterno was a legendary figure in college football. The year before, he'd been inducted into the College Football Hall of Fame. Penn State fans referred to him, reverentially, as "JoePa." By that time, January 2009, he was eighty-two years old. I spotted him in the crowded room immediately—the famous broad face, the thick, old-fashioned brown plastic frame glasses, the sweep of gray hair. I went up to him, introduced myself, and told him that I was committing to Penn State.

Urschel, he said. *From a Jesuit school. I went to a Jesuit myself, back in Brooklyn.* He cracked a joke about the culture of discipline, and then he smiled. *Justice under God.*

Yes sir, I replied.

———————

LATER THAT WEEK, I was in the car with my mom when the phone rang. It was Jim Harbaugh, who was then the head coach at Stanford, saying they'd let the ball drop and wanted to offer me a scholarship.

Stanford, I mouthed to my mom. She started shaking her head vigorously and mouthing *Yes! Yes!*

I had to tell Harbaugh that I had already committed to Penn State. Harbaugh asked how he could change my mind. *I already gave my word*, I said. I thanked him and got off the phone.

My mother was shaking her head, *Stanford!* she cried.

It's too late, I said.

Stanford! she said again, as if she hadn't heard me. But she found peace with it. She had come with me on my recruiting trip and had attended the events for the prospective parents, and she had been impressed by the emphasis on balancing academics and athletics at Penn State. She had looked into the engineering department and satisfied herself that it was strong.

You'll still major in engineering? she asked—though I wasn't quite sure whether it was a question or a statement.

Of course, Ma, I reassured her. If I studied anything else, I'd never hear the end of it.

Sure, I'd become a rocket scientist—why not. But I was going to Penn State to play football.

‖‖

Seven

Arriving at Penn State

Math, 2009–2010

I graduated from Canisius on a Saturday in early May. That night, after partying with my classmates, I packed the car, an old Volkswagen Passat, and my mom and I drove to State College early the next day. The roads from Buffalo snaked through small towns and alongside steep, rocky hills, before entering the endless miles of open farmland in central Pennsylvania. The monotony of the cows and fields gave me plenty of time to think about what I was getting into.

In theory, I came to Penn State to get a college degree, but—like every other recruit—I really came to play football. I wanted to do well academically, but my attitude coming into college was pretty much the same as it had been in high school. As I saw it, I had five years to play football—one redshirting, ineligible for games, and four on the active roster. I had the rest of my life to pursue a career.

My mother, as usual, had a different idea. Every year, she would ask me if I was ready to quit football. *Don't you know how dangerous it is?* she would say. She saw the bruises on my shoulders and around my armpits, the marks left when another player's helmet drove the edge of my pads into my shoulder. She watched from the stands as injured players were loaded onto carts and driven off the field. She taught me how to bandage my cuts with the practiced hands of a trained nurse. *Foolsball,* she'd mutter.

I'm not going to quit, I'd answer, not hiding my annoyance. She'd shake her head in resignation. But she would also hug me. *I'm proud of you,* she would say in her matter-of-fact voice. Then she would flash that wide, vivid smile of hers. *But my baby's going to be a rocket scientist. Don't forget that.*

I meant to follow her advice. I really did. But a few other courses in the department caught my eye first—math classes. I took one on matrices, the rectangular arrays that my dad had shown me in the library in Buffalo back when I was a kid. I took Calculus II, even though I had passed out of the course on the basis of my Advanced Placement score. My academic advisers on the football team tried to dissuade me. *College courses are not like high school classes,* they warned. *They're harder.* But to me, Calculus II was actually easier than high school— and more boring. Vector calculus, on the other hand, was fascinating. (For engineers, a vector is some quantity with a magnitude and a direction in some given space. I imagined a wide receiver tearing down the field: the vector would indicate the direction of his route and the speed at which he was running.) The structure of the course suited me. I was encountering totally new material and was expected to learn it more or less on my own. The professor's style was hands-off. We weren't spoon-fed instructions. It was like being back in my bedroom as a kid, with a book lying open in front of me and no one to tell me what to do. It was up to me to figure it out.

I would wake up early to study, at four-thirty or five in the morning, so that I could get in an hour or two before I had to head to the weight room. The mathematics courses were drawing me in. As the semester went on, I focused less on how mathematics was a description of the physical universe and more on the abstract ideas behind the world. I took mechanics, a physics course that studied objects in motion. Back in my room, while the sky behind my blinds began to lighten into day, I ignored the names of objects in the problems and instead thought of them as abstractions. When I encountered a rocket in a problem, I blurred the image out of my mind and imagined it simply as an object moving in space. Instead of trains and planes, I thought of masses and forces. I took microeconomics too. When I did the work, I focused entirely on the calculus behind it. Money barely figured into it. What interested me were the mathematical dynamics at work: optimization, efficiencies, balances, and distributions. The formulas themselves were often elegant and beautiful in their simplicity, despite the complexity of the human behavior they described.

By THAT FIRST WINTER, I had taken two semesters of classes—and all but one of the required math courses for an engineer. But I had yet to take a single engineering class. My academic adviser, Christine Masters, noticed. A professor in the engineering department, with a tidy wreath of tight brown curls, she had been at Penn State since she had been an undergrad there in the 1980s. She knew how to gently guide wayward students back into line. *Why don't you try engineering science?* she suggested. It was an interdisciplinary honors program within the engineering department, but it had more flexibility than the engineering degree requirements. Dr. Masters was not stupid. She sensed that the way to keep me in engineering would be to let me take as many math classes as I wanted and get as much as possible out of them.

My mother was less than pleased with the switch. *Engineering science? That doesn't sound like aerospace.* I could almost hear her frown over the phone. *It's still engineering, Ma,* I said, trying to appease her. There was a long pause. *Your electives will be in aerospace, right?*

Of course, I said in my most reassuring voice.

Dr. Masters and I agreed upon a plan to let me complete my engineering science degree in three years and begin my master's in engineering in my fourth. As an introduction to engineering, she strongly recommended that I take Engineering Design 100, a course in which students draw engineering designs on the computer via a system called CAD. I refused, imagining that I would be bored by making blueprints—an early sign that, in my heart, I knew engineering might not be for me. Instead, I asked to take a junior-level course in thermodynamics offered by the mechanical engineering department. Masters protested. *You don't have the prerequisites,* she said.

I've gotten A's in all the math classes I've taken, I countered. Plus, I knew from my experience in Calculus II not to listen to the voices that were telling me to slow down.

Fine, she said, sighing. *But on one condition: you have to take an introductory course in statics offered by the engineering mechanics department.*

STATICS IS THE BRANCH of engineering that looks at forces acting on physical systems that are not accelerating. They can be at rest, or their center of gravity can be moving at a constant velocity. Think of a building or, for that matter, a human body standing in the gravitational field of earth. I'd already covered much of that material in high school and in the calculus-based introduction to kinematics I'd taken during the fall of my freshman year, and I was impatient with the pace of the class. I was not always graceful and humble about it either. I took a little too much obvious satisfaction from handing in my exams long before everyone else. The competitor in me struggled not to think of an exam as a race.

Thermodynamics was better, because it was harder. But something about the focus of the course still bothered me. The focus of the course was practical. The laws of thermodynamics describe the interaction of temperature, energy, and entropy, and they help explain how systems change when their environments are altered. They are some of the fundamental laws of physics, critical for any path in engineering. Thermodynamics as a field was developed in the nineteenth century to help make steam engines more efficient, and in the class we were likewise concerned with how the principles of thermodynamics applied to mechanical systems. A formula would appear on the blackboard, and we were supposed to dutifully write it down, commit it to memory, and then do countless computations with it. None of the other students seemed to mind this. It wasn't that they were incurious. They were interested in learning how things worked in order to build and create things. But I wanted to know *why* things worked, not only how. I was less concerned with constructing things than deconstructing them. I wanted to explore the underlying reasons, the mathematical foundations. The only engine I was interested in was the process of mathematical progress. I was starting to think I would be a very frustrated engineer. But what else could I do?

Then, in the spring, I took the last mathematical requirement for engineers, differential equations.

WHEN I FIRST LEARNED to solve equations as a kid (say, $x + 2 = 5$), I generally looked for solutions that were real numbers. If I had two apples and I wanted to know how many more apples I would need to add to my pile in order to have five, I knew that the answer wasn't going to involve apple juice. Once I started doing more complicated math, however, I learned that equations weren't as simple as they first seemed, that there were whole categories of equations and functions— relationships in which each input (let's call it x) would lead to one output—that I had never imagined. (For that matter, there were equations that involved so-called imaginary numbers.) In calculus, I had encountered differential equations, which relate functions to their derivatives (the amount that a function is changing at any given point). Because derivatives essentially represent a rate of change, they have countless applications—including, of course, engineering. They can model processes, describe how those processes alter over time, and make approximations about future states.

Let's say we have a segment of a metal rod, my instructor said one day early in the course, drawing a cylinder on the chalkboard, *and the rod has been heated in one spot by a torch.* I thought about the situation for a second. It seemed pretty straightforward to me. After all, it wasn't unlike sticking one end of a metal rod into a fire. When you took it out of the flame, that end would be extremely hot, but over time the heat would dissipate, moving from the hot end to the cooler regions, until the whole thing was uniformly warm. *How can we represent what is happening mathematically?* my instructor asked the class. That seemed a lot harder to guess. There were things we knew about the rod as he had drawn it on the board—for instance, diameter and length. Because the question involved a process changing over time,

I would have known that it involved differential equations such as the ones I studied in calculus. But it wouldn't be as simple as plugging in a couple of values. There were a lot of variables with complex relationships—the density of the rod, for instance, or whether the two ends of it were insulated—along with the physical laws. Then, there were questions of boundary constraints and dimensions. Differential equations could get complicated quickly.

The heat equation is actually one of the simpler ones, our instructor said after the class struggled through it. Often, differential equations can't be solved explicitly. It can be hard or even impossible to untangle all the variables. Some solutions are vague or even nonexistent, and they require using all sorts of different techniques. Still, differential equations are the most powerful tool we have for describing natural and physical processes.

As THE COURSE PROGRESSED, my mind started moving in a different direction, away from the concerns of engineers. Instead of thinking about physical situations, I started thinking about shapes more abstractly. Visualization was starting to mean something different to me than picturing pipes. Instead of a rod, I would think of a cylinder. I experimented with how to stretch, flatten, bend, contract, and rotate it in my mind. Instead of heat dissipation, I would imagine a more general transformation. Instead of natural processes, I thought in terms of functions. The heat equation could have represented anything. It wasn't the temperature of the rod that aroused my curiosity. It was the equation's intricate elegance and difficulty that spoke to me. Differential equations described something just beyond my understanding— but not entirely out of reach.

The course was taught by a PhD student in the mathematics

department by the name of Yu Qiao. Yu was not particularly striking at first glance. He wore glasses and, most days, a light gray polo. But behind his unassuming appearance, he had a sophisticated mind. Yu took an interest in me. Perhaps it was because I was hard to overlook in a classroom, since I was nearly three times the size of some of my classmates. The fact that I was a football player, of course, made me a curiosity, especially given my strong performance on problem sets. No one singled me out, but still, there was no point in my being shy. I couldn't have blended in even if I'd wanted to.

One day I ran into Yu on a bus, and he began telling me stories about his life. He told me about his childhood in China and about arriving at Penn State as a doctoral student. Yu's adviser had been the head of the department, a famous scholar named John Roe. Roe even had a mathematical structure named after him, a Roe algebra. Yu worked with Roe for several years. But slowly, his enthusiasm turned to a kind of self-doubt, and then the doubt soured into discomfort. *Eventually, I had to admit it,* Yu confessed to me. *I didn't fully understand Roe's work. It wasn't enough to understand it partially. In my dissertation, I was building a house that had no foundation.* Yu switched advisers, and got his PhD a few years later.

Yu's honesty about his difficulties in mathematics left me more impressed with him, not less. It wasn't a sign of weakness for him to admit that, after so much hard work, and after such a long journey, he had reached a dead end. His candor said something to me about the quality of his character. *It is not enough just to love something,* Yu told me. *It is not enough just to pass the exams. You have to choose the right problems and find the right people to work with.* He shook his head. *Mathematics can be unforgiving,* he said. *You can't fake it.*

I knew it was a cautionary tale, but there was something alluring about his warning. You can't fake it: that appealed to me.

One day that spring, we were studying a particular technique to solve differential equations when Yu challenged us with a problem: $4x^5 - 4x^4 + 5x^3 - 5x^2 + x - 1 = 0$. *Find the roots of this equation,* he said with his typical smirk. He knew there was no explicit method we could use to solve this kind of equation (called a fifth-order polynomial—a polynomial because it is an expression of more than one term, and fifth-order because its highest term has an exponent of five). He was sure he had us stumped. It was his intention to let the question sit for roughly thirty seconds, and continue through the exercise. He clearly wanted to put us in our place.

But I did not know that we were not supposed to be able to find an answer in our heads. I thought Yu was asking us a real question. While my classmates stared into space, waiting for Yu to give the solution, my brain was turning. I'd always had an ability to look at polynomials and factor them almost intuitively, ever since middle school, when I'd drawn those brackets and let the numbers tumble out of my mind and into them. This one was of an equation of a much higher order, but it did not strike me as essentially different. And in fact, within a short time, I had the answer. It involved imaginary numbers (i, or the square root of -1). I raised my hand.

Yu called on me, looking confused. His look turned to astonishment as I rattled off the roots: 1, $\pm$ $i/2$, and $\pm$ i.

For a few long seconds, Yu was silent. Finally, he asked, *How did you solve it?*

I just did, I replied. I seriously didn't understand why it should be unusual, even though Yu was clearly taken aback.

Yu knew what I didn't: he knew the Abel-Ruffini theorem, also known as Abel's impossibility theorem. The theorem says that there is no general algebraic solution to polynomial equations of the fifth degree or higher with randomly chosen coefficients. That doesn't mean that there *isn't* a solution. Obviously there is; I figured it out! What it means is that there is no way to do it that you can apply to *any* fifth-order polynomial. I had solved the equation not with some formula I had memorized, but guided by my instincts. He quickly gathered himself and went on with the rest of the lecture.

After class, I stepped outside into the day in a kind of daze. In my memory of that moment, the warm sun seems a little brighter, the campus lawns a little greener. Something had awakened in me. I had never thought that my mathematical ability was anything special, but it was dawning on me that it might be unusual, and that it might in fact be something that defined me. I had always thought of myself as a football player, first and foremost. I knew football wouldn't last forever, but so far I had let my mother worry about what came next. As I walked across the campus, though, my sense of myself shifted. It was a little ironic, maybe, that just as I was fulfilling my football dreams—getting the chance to play for a great college program—I started to think that maybe I might have other dreams worth pursuing. I felt a new desire stirring.

I looked at the other students going on with their days. I felt a new sense of purpose—a calling. But I didn't have time to dwell on the strange new feeling. I had to get to practice.

||||||||||||||||||||||||||||||||||||||||||||||||||||||||

Eight

Pushing until Failure

Football, 2009

I was more nervous than excited when I arrived at Penn State. Part of me couldn't believe that I had gotten the scholarship in the first place—and an even bigger part of me was convinced that I couldn't keep it. Before I had left for State College, my dad had done some research, so I was aware of something that colleges don't like to tell recruits. Penn State had offered twenty-seven incoming freshmen football scholarships, but not all of us would actually be getting those scholarships that first season. The NCAA limits schools to twenty-five free rides per class, and so at least two of us would be "grayshirted"—which meant that our scholarships would be held back a semester, until the spring term, at which point we would become full-time students and members of the team. If a grayshirted football player wants to take classes in the fall, he has to pay his own way. I was the twenty-sixth recruit, and one of seven offensive linemen taken in my class. I knew there was a decent chance that I would be one of those two left on the outside looking in.

The day I arrived in State College, after my high school graduation, was the day before the start of the first summer term. The freshmen on the football team weren't required to arrive until the start of the second summer term, but I was taking advantage of Canisius's early graduation date, in mid-May, to start training with the upper classmen a few weeks before the rest of the freshmen arrived. It was my dad's idea. He thought that I would benefit from extra time working out with the team. I knew he was right, but there were a few disadvantages to the plan too. It meant I was showing up by myself, the only new guy.

Coach McQueary met me and my mom in the parking lot of the football building when we pulled in. *Your first workout is at eight a.m. tomorrow,* he said. *Your ID will let you into the building.* And then he was gone, and I was on my own.

I unpacked and tried to make the empty dorm room seem more like home. I hadn't brought much, though, and so after an hour or so I was left wondering what to do. I had no internet access, no idea where things were on campus. I knew no one. Dinnertime came and passed, but I didn't know where to find food. I was ravenous.

Finally, my roommate, Ty Howle, walked in. Ty was a freshman on the football team, like myself. He was a center from North Carolina— shorter than me but more solid, with a round face and close-cropped brown hair, and a solid chest that suggested power. I had met him for about twenty seconds during my recruiting visit. Ty and I were in the same incoming class, but he was such a highly sought recruit that he'd committed to Penn State the previous March—nine months before I had—and had graduated from high school that fall so that he could start working out with the team that spring. We played some *Madden,* and I started to relax—enough so that I finally got up the courage to ask, *You know where to get some food around here? I haven't had dinner.* Ty cocked his head to look at me, as if sizing me up. *You should have said something sooner. There's a McDonald's close by,* he said in his slow Southern drawl. He stood up. *Let's go.*

Ty was one of the easiest people to be around that I'd ever met— funny without being obnoxious, thoughtful but not too serious. We quickly figured out when to give space, when to give support, and when to give each other a hard time. Within a few weeks, I felt like I'd known him my whole life.

I DIDN'T KNOW WHAT TO EXPECT from the other guys on the team. It mattered too—not just because of my anxiety about fitting in. In football, the team is everything. *Everything.* No other major sport compares. A quarterback, no matter how great, can never win a game by himself. A safety needs his cornerbacks. The center relies on the right and left guards. There has to be enough trust, tight enough bonds, that every player is willing to risk injury so that a teammate can advance just a few more yards down the field. You don't have to really like the guy on your left, but you do have to be willing to sacrifice your body for him.

It also mattered because these were the guys I would be spending the better part of my time with. We would be waking up hours before other students to lift weights and practice. I would eat most of my meals with them, spend summers on campus with them, and travel around the country with them. And if I was really going to be part of the team, I would be going to bars with them, taking classes with them, maybe even eventually sharing an apartment with them. I had very little idea of what to expect. Canisius had not been the kind of high school where a football player would open a locker and find a stash of cash from a booster, or where girls offered themselves to star players. (For starters, there were no girls.) But of course I knew those high schools existed, and it was possible that there would be a few recruits who'd gotten every pass in life because they had pulled in a big catch in the state championship game.

I didn't know what the others would see in me either. I was an undersized recruit, and not prized. I had terrible technique. I was well aware that the team didn't really need me, and I had a small but persistent worry that it wouldn't want me. I could still remember what it had been like as a kid to be the one that no one wanted to be around. I'd never felt that way at Canisius—but I was a long way from Buffalo now.

So I kept quiet. There was only one way I knew how to deal with the uncertainty: control what I could. I could keep my head down. I couldn't turn myself into some outgoing, social person whom everyone was drawn to. But I could work hard. Even if I wasn't automatically liked, I could be respected. So I'd listen calmly when the coaches yelled at me, which they did every day. I'd figure out whom not to mess with and keep my distance. I'd learn who the leaders were.

ONE DAY THAT FIRST WEEK, I was in the weight room lifting when I saw Daryll Clark, the quarterback. Daryll was one of the best quarterbacks in the country, and the team's captain. The season before, he'd broken several Penn State records. He was about my height, African-American, solid and strong, and even in the weight room he had a kind of aura. I stood up a little straighter when I saw him.

I went up to him and introduced myself. *I know who you are,* he said as he looked me over. Then he said, *You want a haircut?* For a second, I wasn't sure whether I had heard him correctly. I stood there nervously, not quite sure what he meant. Did my hair not look right?

Yeah, you want a haircut, he continued, not seeming to mind my silence. *I cut hair. I'll give you a nice, fresh cut.*

Yeah, I could use a haircut, I said, trying to sound casual.

Here's my number. You call when you want to come over, and I'll cut your hair. He turned away and went back to lifting.

I went over to Daryll's place, got my hair cut, and sat down on a fraying couch with a few of our teammates who were hanging out there. In some ways, fitting in was that simple. That was part of the beauty of being on a team: unless you give the other guys a reason to regret accepting you, you belong.

But I wasn't about to take anything for granted—least of all my place on the team. Ty and I would make plans to lift weights on Saturday or Sunday mornings at eight, and I'd wake up at six, get dressed, and sit on my bed, waiting. *I can feel you staring at me,* Ty would mumble, half asleep.

I MANAGED TO AVOID BEING GRAYSHIRTED, but as most freshmen players are, I was redshirted, which meant that I wouldn't start playing in games until my second year in school. That gave me an extra year of eligibility on the other end. Every Friday before a game, there was an extra lifting and training session for those of us who weren't going to play, and the strength and conditioning coaches would use it to try to crush us. There was one trainer for every two players. No one wanted the head strength coach, John Thomas—JT, we called him—to take them through it. JT had been at Penn State for nearly twenty years, but before that he had been the head of the strength and conditioning program at the U.S. Military Academy. Sometimes it seemed like he thought he was still at West Point and we were grunts in his army. He was known for trying to break your will. JT always picked me.

JT made it his personal mission to make me suffer. Sometimes it felt like he was trying to make me quit. He would have me do wall-ball push presses for twenty minutes straight. He'd make me do squats until I couldn't stand.

I was miserable. Every Friday, I arrived at the weight room hoping to avoid JT. And every Friday, there he would be, waiting for me. *Pull-downs until failure*, he said once, adding the maximum weights. I pulled and pulled until my back burned, my arms wouldn't stay raised, and my grip had totally lost its strength. Then he took all the weight off the bar. *Keep going*, he said. I looked at the bar, at once humiliated and at the same time aware that even that weight was more than I could bear. Somehow, I managed to reach up, but I couldn't feel my hands and couldn't hold the bar. He cursed at me. I grunted and kept trying. Finally I managed to hold on to it—but I still couldn't pull it. He kept barking until somehow I managed to bring it down.

I had only one goal that fall, and that was to not complain to JT. I would shake and choke, but I would black out before I said that I couldn't do something. That was pride, mostly, but I also knew that JT's barbaric ways were working. I was getting stronger. I put on thirty pounds of muscle during my freshman year. The soreness I felt every night when I went to bed and every morning when I woke up was a reflection of the way my body was changing.

It wasn't any easier on the field than it was in the weight room. On our first day in pads, I went up against a backup defensive tackle, a guy who was thirty pounds heavier than me—if I weighed myself while wearing my helmet. His girth was actually deceiving: he moved like a much lighter man. I'd never faced someone who got off the line so fast. I barely had time to get my hands up before I was pushed backward, stumbling, and then blown by. The second time was even worse. Before I realized it, I was on my back, trying to catch the breath that had left my body on impact. I turned my head just in time to see the offensive line coach staring at me, before he shook his head, turned, and walked away.

It was very clear, early on, that compared to the other linemen, I was terrible. My technique was still horrible, and I couldn't compensate with strength or size. The only advantage I had was a willingness to suffer more than most—and I could take criticism, which I got a lot of. When I did something wrong, I heard about it. But for some reason, the offensive line coach seemed to like me, and he spent extra time working with me, fixing the way I shifted my weight, where my hands should go. As I got stronger, my technique started to come together. One day that fall, I stoned one of our stars, Jared Odrick—a future first-round draft pick—on the line. He was so mad that he threw his helmet. In that moment, I was convinced that I could play. But there was still a long way to go.

Nine

||||||||||||||||||||||||||||||||||||||||||||||||||||||||||||||||

Probability

Math, 2010

During my freshman year, I thought a lot about luck. I knew that a lucky string of events had brought me to Penn State. I also knew that an unlucky sequence could have just as easily sent me somewhere else. It was inevitable that players who were far more naturally talented than I was were overlooked, for whatever reason, or chose another school when they could have taken a spot at Penn State. So I very consciously tried to identify what I could control and what I couldn't, and I tried to focus my energy on what I could. I couldn't always make my own opportunities, but I could try to put myself in a position to take advantage of them when they came. Some people, I know, can't turn their anxiety off. When it came to events that were out of my control, I could. Of course I got nervous and worried—but I learned to step back and think about what worries were valid and what were not. If I couldn't change the outcome, then I'd let the worry go, or at least shove it to a place inside my mind where it couldn't affect me daily. Long before I was particularly good at anything, I was an elite compartmentalizer. It didn't make sense to worry about what wasn't in my power, or to fight battles I couldn't win. Luck would influence my life, no matter how much I planned or how hard I tried.

Sometimes I thought of that sequence of events that brought me to Penn State—McQueary's unexpected visit, the timing of Stanford's call. At the same time, I tried not to put too much emphasis on coincidences, even when other people found considerable meaning in them. For example, whenever I walked into a party, I knew that there was a very good chance that two people there shared a birthday (if not necessarily the same year). If there were twenty-three people there, it would be around 50 percent. And if the party was big enough—around seventy people or more—then I knew that it was almost certain that two people were born on the same day of the same month. *How is that possible?* people would ask me when I said this. *There are 365 days in a year.*

It was easy to explain the math to them—even the mathphobic nonbelievers. In fact, it's a famous problem, one of my favorites—and it's not a bad party trick. If I know the probability that *no* two people share a birthday, then I also know the probability that two people do. Since those two events are mutually exclusive, they add up to 100 percent. The chance that the twenty-three people all have different birthdays can be thought of as the probability that the first person in a group doesn't have the same birthday as the second, and that the third person doesn't have the same birthday as the first or second, and so on. That's easy to calculate: the probability that the first person and second person do not share a birthday is 364 divided by 365. The chance that the third person doesn't share a birthday with either of the first two, given that the first two don't share a birthday with each other, is

363/365. I could keep going until I figured out the probability that the twenty-third person has a different birthday from all the rest, which is 343/365. To calculate the probability that all those events will happen, I would multiply all these individual probabilities together. That gives me .492703. Subtract that from 1 to get the probability that two people *do* share the same birthday, and you get 0.50729, or just over 50 percent. The important thing here is that the birth date isn't specified. The chance that two people share a birthday on, say, April 22 is much lower. A more general way of saying this is that improbable things happen all the time in life. It's just extremely hard to predict *which* improbable things will occur.

Sometimes, it seems like random events are beyond the scope of human understanding. I can't explain the chain of events that brought me to Penn State. I can understand why people talk about fate or turn to God for explanations, or shrug as if randomness is totally beyond human comprehension. But a lot of randomness can in fact be measured.

During the summer term after my freshman year, I decided to take a more serious math class, Advanced Probability—a class that would turn out to influence my thinking in all areas of my life, though I wasn't aiming for that at the time. As I grew older and more accountable for my choices, I became more aware that understanding how probability worked helped me make better decisions. Every day, I weigh the likelihood of outcomes, big and small, usually unconsciously. Very rarely does it involve calculating numbers. People don't literally compute probabilities when they decide where to live, what job to pursue, or whether to take a daily aspirin to lower the risk of a heart attack. But people are prone to overestimating the chances of certain things and underestimating the chances of others. They fear terrorist attacks on airplanes but don't think twice about driving on the freeway. (You're more likely to die in the bathtub than from a terrorist attack!) They have intuitions about what's likely and what's not, but those intuitions are often off.

When I had started learning about basic probability as a kid doing puzzles—long before I knew what probability theory was—I'd been a little unnerved to realize that my intuition could sometimes be so wrong. I was a person of strong instincts, and those instincts served me well. I trusted them. So I had been startled when I'd come across brain teasers and puzzles that had counterintuitive results. I wanted to have better intuition.

One of the choices I was making as I entered my sophomore year was whether to remain in engineering or become a math major. I had a strong sense that I would be happier in mathematics, but that wasn't based on taking a lot of classes in the mathematics department. I still wasn't sure what a mathematician did, and I could guess that it wasn't sitting around and doing puzzles all day. In fact, I wasn't really sure what mathematics *was*.

Like most people, I'd taken it for granted that mathematics dealt in certainties. Traditionally, mathematics is supposed to get rid of uncertainty, not account for it. A mathematical statement is true or false. There is no *maybe* in an equation.

But I was also becoming more and more alert to the ways in which math described the world, and I knew from my own experience that randomness governed so many events. There were a lot of *maybes* in life.

Probability theory is math's answer to uncertainty: a way of dealing with random variables and events. The outcome of a single trial can never be predicted with perfect certainty, but probability theory allows us to describe and quantify the possible outcomes. It supplies tools to create models that account for the sources of randomness and measure our doubt. It's most commonly talked about in terms of gambling, but it has applications in computer science, finance, physics, engineering, the military, meteorology—any profession that involves interpreting data or modeling events influenced by randomness and chance. And it has huge, and often unrealized, implications in daily life. Later on, I would appreciate how probability helped me understand and evaluate my choices. At the time, though, I was focused on one thing: I needed to understand probability if I was going to pursue math.

When I signed up for the probability course, my advisers tried to talk me out of it. *You can't take that class*, my football advisers told me. It was a course for seniors; I was not yet even a sophomore. I wasn't getting any help from within the math department either. *You don't have the prerequisites*, I was told yet again.

Nothing made me want to take that class more than being told that I couldn't do it. That was my nature: if someone thought I wasn't up for something, then I had to show them wrong.

Finally, I persuaded my advisers and the math department to take the risk. The special-permission forms were signed. It was a gamble, from their perspective. From mine, it was a necessary test. I wanted to know what I was capable of. I liked to bet on myself.

PROBABILITY HAS ITS ROOTS IN GAMBLING. Given how long people have been betting on games of chance such as cards and dice (and given your average casino scene, you cannot convince me that the cave men did not), I would have guessed that basic probability theory would have been around almost as long as arithmetic has been. I was surprised to learn that it wasn't made rigorous until the seventeenth century, after two mathematicians, Blaise Pascal and Pierre de Fermat, received a letter from a gambling friend asking for some advice. He needed to figure out how to split the pot when a game was interrupted with one player ahead. How could they quantify the chances that the losing player might have had the luck to come back to win? The idea of odds was already familiar; even their gambling friend knew that if he rolled a die, he had a one-in-six chance of rolling a six. But just how those odds worked, and how to use them to predict future events, had never before been formally worked out or even well understood. After Fermat calculated the probability of all the possible events, Pascal came up with the elegant idea of weighting the average possible outcomes by their probability. Their exchange revolutionized math—and enraged their friend, who had assumed that all you had to do to calculate the odds that you would roll at least one six in four tries was multiply $\frac{1}{6}$ by 4, which would give you $\frac{2}{3}$. Pretty good odds! When Pascal explained to him that, in fact, it was just over 50 percent (or $1 - (\frac{5}{6})^4$, where $\frac{5}{6}$ is the chance of *not* rolling a 6), the friend said "that arithmetic was demented."

Other mathematicians heard about what they were doing and started circulating their work. It was immediately obvious that their methods opened up a whole new area in mathematics. Pascal famously even used probability to argue for believing in the existence of God.

One of the most interesting and useful results was worked out quickly. It's called the law of large numbers. People had already observed that when you repeat an experiment, the results will vary, but if you do it enough times, the outcomes will tend toward some result. Think of flipping a coin: if you flip a quarter ten times, you're unlikely to get exactly five heads and five tails. If you flip a coin a hundred times, the chances that exactly half of those flips will turn up heads is incredibly small. If you were to flip a coin 10,000 times, the chance that exactly 50 percent of those flips would be heads would be almost zero, but it's very likely that the ratio of heads to tails would be closer to 50 percent than that of ten flips—and if you could flip it an infinite number of times, it would approach 50–50 with probability 1.

In the early eighteenth century, Jacob Bernoulli gave a precise mathematical framework for showing that the proportion will almost surely converge to a fixed probability as the number of trials increases. It's easy to misconstrue what that means—and misunderstanding it has led to a lot of bad decisions. Bernoulli was not saying that you can predict the outcome of any one trial with certainty. The law of large numbers does not mean that if you flip a coin ten times and it lands heads every time (which is entirely possible!) then you're more likely to get tails on the eleventh flip. (This is known as the gambler's fallacy.) You are just as likely to get heads on the eleventh flip as you are on the first. Every trial is independent, not connected to what has come before. What the law of large numbers *does* mean is that if you flip a coin enough times, then those first ten flips will be overwhelmed by the results of all the other flips taken together.

I already knew the value of being comfortable with concepts such as the law of large numbers—for everyone, not just for mathematicians. But in probability class, I encountered ideas that challenged the way I thought about math—and influenced the way I thought about life. From time to time, I found myself thinking about one problem in particular. There are infinitely many decimal numbers between 0 and 1. Suppose a number is randomly generated. Each number—.1, .434342, .8382948398234, and so on, infinitely—has an equal probability of being chosen. So how do we determine what that probability is? It turns out to equal zero. There is a zero probability of choosing .1, or .434342, or .8382948398234, or any other number. (Note that if a human were the one choosing the number, this would not be true.

Humans are very bad at doing things randomly.) And yet it is possible. After all, *some* number is chosen. A zero probability event is not the same as an impossible one.

There are formal ways of showing this. In class one day, my teacher mentioned the man who came up with a way to talk about these kinds of probabilistic questions: Andrey Nikolaevich Kolmogorov. Kolmogorov's main insight was that, from a mathematical perspective, probability is simply the measurement of the likelihood of an outcome. He already knew of a technique called measure theory, which generalized intuitive concepts such as length, area, and volume. So he managed to use measure theory to generalize probability, placing it on a rigorous ground.

When he did that, he did something more than let us answer tricky questions like that one about the probability of choosing a number between 0 and 1. He formalized the field. How he did it made an impression on me. Kolmogorov established a series of rigorous definitions and axioms, or self-evident principles, that absolutely had to be satisfied in order to accept something as true. Before him, probability theory had been vague and even a little random. He built probability theory from the ground up. It allowed mathematicians to talk about uncertainty with certainty.

I felt a little like Kolmogorov in that class, building my understanding level by level. My adviser was right, I thought. When the class began, I was a little lost. I didn't have the right base of knowledge. I didn't have the prerequisites. So I had to establish the foundations myself. I took nothing for granted. I had to absolutely understand each step before I could move on.

Ten

||||||||||||||||||||||||||||||||||||||||||||||||||||||||||||

We Are

Football, 2010

That summer, after my probability course was over and training camp began, I earned a spot on the offensive line's second string. There were three players backing up the interior—two guards and a center—and I was the youngest of the three, and the lowest on the depth chart. For most of my redshirt-freshman season—my sophomore year as an academic student—I never got on the field except for garbage time.

Late in the season, our backup center caught mononucleosis. Ty Howle, my old roommate (and my best friend) took his spot. Then, the night before we were set to play Indiana at the Redskins' stadium outside Washington D.C., our other backup guard was late to dinner, and so Paterno declared that he wasn't allowed to play in the game. (*If you're five minutes early, you're late*, Paterno liked to say.)

Then, early in the game, I saw our starting center go down.

I yelled at Ty to get his helmet. So did one of our coaches, who was on the field with the injured player. Ty started to run out onto the field. But Paterno stopped him. *Urschel!* he yelled. *Where's Urschel? Move Wisniewski to center and put Urschel at guard.* I grabbed my helmet and ran onto the field, my heart beating fast.

We broke a tie in the third quarter and ran away with the game, 41–24. I played well—very well. The offensive line gave our quarterback time to throw and opened gaping holes for our running back to race through. We finished the game with 496 total yards of offense.

My performance caught the coaches' attention. It happened at just the right moment, as their minds were turning toward the following season. I'd earned a serious look as a potential starter. But I knew that my merit only carried me so far. I was lucky. I had been given the chance to prove myself then only because a player came down with mono, another was late (a very rare occurrence in Paterno's world), a third player went down with an injury, and then our backup center was stopped as he ran onto the field. If any one of those events had not happened the way it did, I'm not sure what my football career would have been. Maybe I would have gotten a chance to play in a meaningful game a few weeks later or the following season—or maybe not. There are a lot of people in this country who are talented and work extremely hard, and for one reason or another, they never get the chance to prove it. For a string of random reasons, I did. I knew what a small chance there was of all those events happening in sequence. But I also knew that improbable events happened frequently.

There wasn't time, anyway, to worry about what I did and didn't deserve. As I became more serious about math, my life began to bifurcate. My academics and athletics were bound only by the discipline and competitiveness that I brought to both. My advisers stopped complaining about the courses I'd signed up for. I didn't give them any choice. I was getting A's in the classroom and improving on the field as well. Sometimes my teammates and classmates and even strangers who had heard about a mathematics major on the football team would ask me how I managed to do both. It was hard to satisfy them with the true answer: I didn't waste energy worrying about whether I was trying to do too much.

I loved both math and football. That made it easier for me to prioritize them above other things. It helped that I had the right temperament. I could focus on my work so intently that I couldn't hear a person speaking to me from only a few feet away. And my unusual ability to compartmentalize came in handy. When I left practice and sat down with a math book, I stopped thinking about whether I had missed a block, or whether my hand was hurting, or anything else having to do with football. Likewise, when I got to practice, I completely forgot about whatever math problem I had been obsessing over only minutes before. When I was focused on one thing, I thought of nothing else.

It was easy for me to move between football and math. I imagine it is not unlike the way a bilingual person moves between languages, or the way a person at home in two cultures can live in both. Math and football were simply two distinct parts of myself. The drive to do them both well came naturally to me. It took some good time-management skills, but there is more time in the day than most people usually take advantage of. College students spend a lot of time hanging out, going to parties, doing extracurricular activities, cultivating relationships. I had two clear priorities—studying math and playing football—and everything else was secondary. That's not to say that I was a complete hermit, and my friends would make sure that I sometimes got out. Most days, though, I would wake up at five or six and read math books before heading to the training room to lift, or the field for practice, or the facility for meetings. At night, I would work until I was too exhausted to do anything but sleep.

The ability to do math came much more easily than football did. I had to work much harder to keep up with my teammates on the football field than I did to keep up with my classmates in math, no matter how hard the course. In high school, I had tried to deny it and wished it were otherwise, but on some level I had always known it. In college, I accepted it. I wasn't an extraordinarily talented athlete, but I did have certain intellectual gifts that not many other students had. I could hold a problem in my head for a long time, probing and testing it, trying new techniques. I had good mathematical intuition, a kind of extra sense that guided me. I had visualization skills that let me spin, stretch, and contract shapes. I was quickly comfortable working with abstractions, even in higher dimensions that were hard for most people to imagine.

Football was different. I was surrounded by guys who were bigger and stronger and quicker than I was, who had natural talents that surpassed my own. It could be frustrating sometimes. I'd come home from practice having been beaten and embarrassed, and in a fair amount of physical pain. I made a lot of mistakes. Just because I was good at math and had a decent memory did not mean that I was any better at executing the playbook than the other guys, no matter what kinds of grades they were getting in class. Some of them had been playing football since they were children. Football was what they knew; they'd known it all their lives.

I had a few advantages, though. I was as competitive as anyone—if not more. My work ethic was strong, and I was able to learn from my mistakes. What also helped is that I knew I didn't have a choice: I *had* to work as hard as I could, or I would lose my spot. I could get by in class if I skipped a lecture or an assignment. If I slacked at training, on the other hand, I thought I might never be able to catch up. But that was one of those worries that I could do something about. How hard I worked was in my control, so I didn't let up.

My teammates and friends—and sometimes reporters from the student paper—asked me if there were any connections between math and football. Sometimes I'd try to give the answers they wanted. I'd talk about problem solving, or I'd talk about the basic physics of being an offensive lineman, using forces and leverage to overpower my opponents.

But in fact, on the field, I never thought that way at all. I simply moved. That was the whole point of practice: to train my body to act before my mind could get in the way. Through countless hours of breaking down the mechanics of a motion and training my muscle memory, I developed a physical intuition that would guide me much better than my conscious mind ever could. If I took one millisecond to consider forces and mass, I would be run over by a defensive tackle— or worse.

That was part of what I loved about football. It was a way of becoming a part of something bigger by losing myself in it. Every week that the team played at home, I looked forward to the moment on Saturday mornings when the bus drove from the team hotel to Beaver Stadium past the thousands of fans lining the streets. They carried signs, wore the school's blue and white, and roared as we rode past. Just before the game, when we ran out of the tunnel into the stadium, I would look out at the sea of supporters and let the roaring waves wash over me.

When I was on the sidelines, I could watch the game like a fan. But on the field, when the ball was live, I couldn't see the people in the crowd, could barely hear them. If it was raining, I couldn't feel the water lashing my bare skin. If it was cold—and, for months, it would be very cold—then I was numb to it. All I could feel was the adrenaline pumping through me. I was aware of the quarterback behind me, the center to my left, the tackle to my right, the row of men in front of me, and my determination to push them back and beat them down. That was it. There was something animalistic about it.

I loved how simple it was. White helmet, blue stripe. Blue jersey, white numbers. No logos. Black shoes. We would yell *WE ARE*, and the response was *PENN STATE*, and I swear, it felt oddly profound. *We are.* That was enough. Nothing less, nothing more.

THE STRANGE THING WAS that the more confident I became as a member of the Penn State football team on the field, the more confident I became of being myself off of it. Sometimes the guys would make a joke about the level of their intelligence or make fun of mine, but by then I knew the difference between the kind of mocking that was mean-spirited and mocking that meant you were just one of the guys.

When I sent my TV home so that I'd have fewer distractions, the news spread through the locker room like a good piece of gossip. My teammates considered holding an intervention. Later, once I started doing mathematical research, I sometimes would lock myself in an empty classroom for days at a time. One of my teammates would come by to check on me regularly, to make sure I had food and hadn't gone *A Beautiful Mind* on them. They all thought I was a little strange, but I also thought I was a little strange. That was okay.

Football has a reputation for being conformist, and there are good reasons for that. It can be. I was comfortable with conformity. As a kid, I'd embraced the idea of doing what the others did, and it helped me fit in. But in college, at least at Penn State, the guys are pretty laissez-faire. You do your own thing, and as long as you're not a jerk to the others, no one really cares. There were those who enjoyed the social prestige of being a football player and all the benefits that came with it. But within the team, there was a lot of tolerance for being different. As long as you didn't make your individual choices a problem for the team, there were no problems. Some guys were devoutly religious. My close friend Miles, another offensive lineman, was passionate about bow-and-arrow hunting. Mike, a cornerback, was passionate about making music. Math was just my thing.

For the most part, my interest in math didn't come up, but there were a few exceptions. Sometimes one of the players would text me with a math question, or a question from his girlfriend. Ty once even had me on the phone with his girlfriend's sister, helping her with her homework. Another time, Glenn, a linebacker and one of my closest friends, asked me for help with an extra-credit problem in his stats class. *Sure, come over,* I said. Glenn was a great guy, easygoing and smart, and it turned out to be a pretty interesting problem, so I figured we should really explore it. I took him through every possible approach, probing the question this way and that. At one point, I pulled out my laptop and suggested we try writing some code. He stared at me. *What kind of epic wormhole are we in?* he asked. After about three hours, he begged me to let him go. We'd come up with the answer to the extra-credit question long ago, and I was still having fun playing around with it. *My brain hurts,* he said. *This is a new kind of brain injury.*

In exchange, Glenn and Ty and my other friends taught me to have fun. There were times when I'd find myself in a group or at a crowded bar and the anxiety that I had felt as a kid in social settings would grip me. But my closest friends, like Ty, knew how to pull me out of it, to put me at ease and make me laugh. There were pranks and inside jokes and video games and conversations about girls, nights at bars and entire days spent at the movie theater where a ticket was a dollar. While doing some math at Starbucks one day, I texted Mike, the cornerback, that there was a beautiful girl I was trying to get the courage to talk to. *Don't move*, he said. Twelve minutes later, he was there to wingman me. I had close friends from outside the football team too: a women's basketball player, a swimmer, a couple of students who were studying to be athletic trainers. There were math majors and soccer players and people who just liked having a good time. I often had too much work to go out at night—I knew my priorities—but when I did, we would close down the bars, singing "Piano Man" at last call. I was happier than I'd ever been.

I felt at home. I've never been one to spend too much time looking at the scenery—I'm the kind of guy who likes to keep the shades drawn—but I came to appreciate the way the long slope of Mount Nittany would slowly appear through the weight room windows as day broke. When it was warm enough, I would sit on a bench on the main part of campus with my work and idly watch people walk by, or camp out at a table in the student center, taking breaks from my work for a cup of coffee and to observe the interactions around me. I liked doing math in cafés and busy places, being solitary in the middle of a crowd.

AFTER EXAMS AT THE END of the fall semester of my sophomore (or redshirt-freshman) year, the football team left for Tampa, where we would be playing the University of Florida in the Outback Bowl on New Year's Day. We would practice every morning at a local high school, and then the bus would take us back to the hotel. Parents were there, girlfriends, a big contingent of Penn State fans, and the press, which was even more persistent than usual because of rumors that it might be Paterno's last game. For me, it felt more like a beginning. If I played well in spring ball, there was a solid chance that I would get more regular playing time the following season. I was ready. But that wasn't the only thing I had going on.

Most of the guys spent the afternoons and evenings at the mall or the club, or at events sponsored by Outback and parties on the beach. The leafless trees and cold gray skies of State College in winter seemed very, very far away.

I spent my afternoons and evenings back at the hotel, at the desk in my hotel room or sitting on one of the couches in the lobby, reading books about celestial mechanics. I'd print out papers from the hotel computer to read on the bus on the way to and from practice.

Hey Ursch! the guys would say as they passed me in the lobby on their way to the beach, towel in hand. *Finals are over! You're on vacation!*

This is my vacation, I'd say, holding up my book.

It became a team-wide joke. As usual, the guys couldn't get enough of giving me a hard time. *Only you, Ursch*, my teammates would rib me, shaking their heads. *Only you.*

You're gonna take over the world, Mike, my roommate on the trip, would say as he left me in our hotel room on his way to the beach.

Nah, bruh, I'd say. *I have better things to do.* I wasn't so interested in conquering the world. But there was an asteroid that I had my eye on.

||||||||||||||||||||||||||||||||||||||||||||||||||||||||||||

Eleven

Q.E.D.

Math, 2010–2011

When I decided to switch majors from engineering to mathematics, I justified it to my mother—and myself—by saying that it would give me the flexibility to go into any number of fields after college. I could become a computer programmer, or a physicist, or get a job in finance. It hadn't occurred to me that I might want to become a mathematician.

Then, during the fall of my sophomore year, I took an advanced mathematics course called Real Analysis. The class was taught by a brilliant mathematician named Vadim Kaloshin. Professor Kaloshin is one of the top experts in the world on dynamical systems, which are systems that change states over time. He had a compact build, a full head of thick, dark-brown hair, a broad, open face, and a vaguely distracted air. He had a captivating intensity, a habit of bouncing when he erased the board. His lectures were a little like dynamical systems themselves. They tended to begin in a straightforward manner, covering a subject or going through a proof, but over the course of the class they would take sudden intellectual leaps, becoming something more profound. A proof of a theorem that looked easy enough would turn into a mind-bending lesson on the different sizes of infinities. A question about axioms would lead me to think about the very nature of mathematical truth. I was fascinated.

As I was walking out of class one day, Kaloshin stopped me and asked me to stop by his office. Extremely nervous, I went. I had no idea what it was that he wanted to see me about. I racked my brain, trying to figure out what I had done that could have landed me in trouble. Outside the door to his office, I hesitated, but he gestured me in.

I sat down. *I've been impressed with your work in class*, he said. *I thought you may be interested in expanding your studies.* He took a massive book off his desk and handed it to me. It was *Chaos: An Introduction to Dynamical Systems*, by Kathleen T. Alligood, Tim D. Sauer, and James A. Yorke.

There's a challenge at the end of the first chapter, Professor Kaloshin said. *Period three implies chaos. Can you prove it?*

I waited until I was home to flip through the book. On the title page, there was an inscription from one of the authors, Jim Yorke, dated January 16, 1997. *"Dear Vadim,"* it read, *"We live in chaos. Now, here is the manual!"* I started to read.

The book was dense with equations and strange terms. "Lyapunov dimension." "Saddle-node and period-doubling bifurcations." "Hopf bifurcation." I had taken only elementary physics and a few advanced math classes. This was an advanced textbook. I was going to have to start at the very beginning—not the beginning of the book, but of the topic. It suddenly occurred to me that I did not even really know what "chaos" meant, let alone know what it might imply. I knew how chaos was used in common speech, but I had learned enough to realize that the colloquial definition of a word is often not the same as its precise meaning in mathematics. I could guess that it meant something more rigorous than a state of disorder and confusion. But what?

For the next week, I spent every spare moment studying the basics of dynamical systems (events such as the swing of a pendulum, the turbulence of water, and the trajectory of a planet in space), trying to learn the rules that describe a way a system changes over time. I read while lying in bed early in the morning and late at night, while eating meals, while waiting in lines—basically at any moment I wasn't in class, on the practice field, or studying tapes or my playbook. It wasn't enough to understand terms like "instability" and "stability" or "period sinks and sources" conceptually. I also had to be able to define them mathematically, and then to understand how those definitions interacted. Sometimes I had to refer to more basic textbooks and papers than the one Kaloshin had given me. I was starting from the bottom. After a week of steady work, I was able to produce the proof.

I brought my result to Kaloshin at his office hours and started to explain how I had done it. He stopped me after a couple of sentences with a wave of his hand. *Good*, he said. Startled, I realized that he wasn't actually interested in my answer. The exercise was a kind of test. He simply wanted to hear me tell him enough to satisfy himself that I had done it.

Keep reading the book, he said. Then he suggested another book, *Mathematical Methods of Classical Mechanics*, by V. I. Arnold. I checked it out of the library, started reading it, and then immediately went back to the library to get a few more basic books, so that I could try to understand Arnold's. I had an extraordinary amount of catching up to do, just to be able to follow along with the advanced text. Still, I persevered through it, and came back to Kaloshin ready to discuss it. Then he started emailing me pdfs of published academic papers so that I could read more about celestial mechanics. I had to stop every few sentences to look up unfamiliar terminology or mathematical concepts—and, even after reading some of the papers two or three times, I had to admit to him that I wasn't sure that I understood everything. There was no way to pretend that I had. A single probing question would have exposed me. I remembered Yu's words: *You can't fake it.* Nor did I want to pretend. I wasn't trying to impress Kaloshin. I just wanted to learn as much as I could, as fast as I could. Kaloshin patiently answered my questions. He was very quick to tell me when I was wrong—and direct about why. I liked that. Some students, I knew, hated to be criticized. It made them feel like failures and sent them spiraling, as if a few harsh words could endanger their future job prospects. But I was used to the blunt manner of football coaches. The only way I knew how to improve was to hear an honest assessment of my progress.

Despite his corrections, he was encouraging. I was a sophomore, he reminded me, attempting work that many PhDs struggle with. Some confusion was to be expected. He was struck by the diligence and determination that I brought to the work—the same diligence and determination that I brought to football. *It is fascinating to see your progress and enthusiasm,* he emailed me one day. No other comment or kind of praise could have made me feel so good—or make me want to work even harder.

After a few weeks of guiding my reading on the basics of classical mechanics and dynamical systems, Kaloshin started to give me papers related to a specific problem, known as the three-body problem (technically called the restricted circular three-body problem). *I have two graduate students who were working on a problem, but one of them has given it up,* Kaloshin said to me one day. *Would you like to take it on?*

I did not bother to hide my excitement when I said yes.

Over winter break, while I was in Tampa for the bowl game, Kaloshin outlined the goal of the project in a series of emails. We were going to try to find an asteroid in the main asteroid belt between Mars and Jupiter whose path could change, crossing into the orbit of Mars. It sounded like an astrophysics problem, the kind of question that would require telescopes and spectrographs to answer. But in fact it was very much a math problem. The asteroid and Mars were, in these cases, objects that behaved in certain ways under certain conditions, which could be described mathematically. We could have referred to them as x and y. The problem that Kaloshin proposed was related to a much older general problem, one that had confounded mathematicians and physicists ever since the seventeenth century when Isaac Newton had mathematically described the movement of objects. Is the solar system stable? Or is it chaotic?

I wasn't ready to tackle Kaloshin's problem quite yet. I did not have all the tools I needed to understand the basics of dynamical systems. So Kaloshin led me through it, giving me exercises that taught me how to do things like create different kinds of maps—not maps like the ones we use to get from city to city, but functions that describe the evolution of a dynamical system. *Let me send you a couple more pieces of the puzzle,* he would say in an email, and then he would outline various ways of looking at different concepts, attaching chapters of textbooks for further reading. He told me to teach myself LaTeX, the computer programming system that mathematicians use to write papers. And, after a couple of weeks, when I was ready, he sent me the work that he and a University of Maryland graduate student, Joseph Galante, had been doing on the three-body problem, and gave me exercises that would help me understand the methods they were using. Mostly, though, it was up to me to teach myself—which is, of course, how I liked it. The challenge for me wasn't just to catch up and to familiarize myself with the material they already knew. I had to make myself useful, to learn methods that they had overlooked or were less familiar with. I wanted— and was expected—to make a contribution.

As I read up on the three-body problem, I encountered some of its long history. When Isaac Newton was developing calculus, he had no difficulty in calculating how two moving objects would interact (say, the earth and the sun), but he could not work out the paths of three interacting moving objects (say, the earth and the sun and the moon). For hundreds of years, neither could anyone else. It was considered *the* major problem in celestial mechanics, the branch of astronomy that deals with the movement of astronomical objects such as planets and asteroids and stars. Newton had identified gravity as the force pulling and pushing planets, and another seventeenth-century scientist, the German Johannes Kepler, had identified the path of a planet around the sun (generally an ellipse). Predicting the motion of three planetary bodies, however, required many simplifying assumptions and a long list of interrelated equations. But that did not deter most physicists and mathematicians from assuming that there *were* ways to predict the motion of celestial bodies precisely. They believed that the universe was complex but also that it was like clockwork, designed by God and governed by physics, moving according to perfect and predictable laws. Until those laws were revealed, mathematicians worked with approximate solutions and made more limited discoveries. They could talk about certain well-known points in an orbit, or families of solutions. But a true understanding eluded them.

Then a Frenchman named Henri Poincaré started exploring the problem. Poincaré was not only one of the leading physicists and mathematicians of his era but also one of the leading figures in all of France, the chief engineer of the Corps des Mines and later the president of the French Academy of Science. He was also a defender of the role of intuition in mathematics, which was under attack at the time. Intuition, Poincaré wrote, was the way through which "the mathematical world remains in contact with the real world; and even though pure mathematics could do without it, it is always necessary to come back to intuition to bridge the abyss [that] separates symbol from reality." That was exactly what I believed. Intuition connected the real world with the ideal one.

As I talked to Kaloshin and read more of the work of other mathematicians, I stared to develop a better sense of what mathematicians did and why it mattered. It is too simple and reductive to say that the work they did was often "useful" in a direct sense. Their research was not the kind that most other scientists did. It did not involve experiments conducted in laboratories. The experiments they ran were in the mind, and the data was more likely to be theoretical than recorded by an apparatus. The tools they used were not telescopes and microscopes but mathematical constructs. Their work was important regardless of whether an astronomer might use it to collect better observational data about a particular asteroid. It was, in fact, highly theoretical—as was the problem I was working on with Kaloshin. Still, the work that they did had significance in the real world. Mathematicians developed techniques and ideas that advanced knowledge. They were able to illuminate something about mathematics or the sciences, and so of the world itself. Intuition was, as Poincaré wrote, the bridge that kept the ideal world in contact with the real. It was what allowed me to look at a wobbly circle drawn by my own hand and see it as an imperfect reflection of the ideal shape.

Poincaré did not want to depend on the imperfect data collected by astronomers or the complicated equations that were jury-rigged together to come up with an approximation of a planet's path. He wanted something precise and conceptual. He was interested in physical systems and what they looked like, how they worked and why. Poincaré wanted insight. He wanted a picture—not the kind of lopsided picture that observational data produced, but the ideal picture behind the imperfect reflection.

The imaginative leap that he made is the kind I admired. Instead of simply refining the way that scientists had used to approach the three-body problem before, he re-conceptualized it. Instead of studying the trajectories of three bodies moving in three-dimensional space (which is extremely difficult), he realized that you could study the points at which the trajectory of the asteroid passed through a two-dimensional plane. Imagine a screen placed perpendicularly to the orbit of an asteroid. Every time the asteroid came around, it would pass through the screen, leaving a hole. Poincaré noted the coordinates at each piercing. If the asteroid was traveling along the same path at the same speed, it would pass through at the same spot every time. But an asteroid did not have to have such a simple orbit. It might move in a way that seemed unpredictable for a long time, and so make a long series of holes. The resulting picture is known as a Poincaré map.

If the asteroid gradually tended toward an orbit that moved through the same hole every time, Poincaré concluded, then its orbit was stable. If an asteroid had been passing through one point but then began to veer away, then its orbit was unstable. By studying the plane map, Poincaré identified special curves that were stable or unstable. He figured that if an orbit moved away from one point, it would eventually settle into another. That was consistent with his (and pretty much everybody's) belief that the universe itself was stable.

Then someone pointed out an error in his original paper. Here is the second part of Poincaré's approach that fascinates me: instead of giving up, or trying to force his earlier conclusions to account for the error, he questioned his original assumptions. He accepted his error and let himself be open to the unexpected. Then he saw it: it turned out that very strange things happened where the stable axis and the unstable axis crossed. Basically, any asteroid on or near the unstable axis would move in a way that was crazy—all over the map. It was chaotic.

I DEVOTED ALL THE EXTRA TIME I had during the offseason to working on the research. I emailed Kaloshin every few days with updates of my progress, telling him what I was reading, and asking questions about concepts when I wasn't quite sure. He would engage with my ideas, offering his own suggestions and thoughts, and guide me toward answers instead of simply giving them. *I have the idea of the twist condition, but do not yet completely understand how, given an arbitrary map, or this map in particular, to show if it satisfies the twist condition,* I wrote in a typical exchange. *I know that it means that given any "vertical" line, the map monotonically twists it and that it involves the partial differentiation of the Hamiltonian, but I'm not 100 percent sure on how to show it. But with more reading I'm sure I will get a better understanding of how to show this.*

He responded:

> *At this point, let me introduce you to a paper that suggests a method of destructing invariant curves. I suggest to split this part into two.*
>
> > *1. How to obtain a (Poincare) map for the RPC3BP?*
> > *2. Suppose you have a (Poincare) map. How to rule out invariant curves?*
>
> *The attached paper deals with the latter issue. . . . It might be directly useful for our project.*

I wrote back:

> *As far as the Jacobi constant goes, my intuition tells me that as the constant increases, the Hill regions decrease, which would result in less and less chaos, since the possible motions of the orbit become less and less. . . . But that is my initial thought. Let me know if this is at all reasonable.*

He replied:

> *It is reasonable and the right line of thoughts. However, it is not totally convincing. It might happen that a Hill region gets smaller, but the orbit we are interested in still passes closer to Jupiter.*

Before long, he and Joseph Galante, the Maryland graduate student, were asking me for my thoughts on various methods and to research new approaches. In January, over the winter break, Joseph came up to State College for a few days, and we mapped out a plan to approach the research. I executed much of the work, as Joseph gave me direction, with Kaloshin weighing in frequently as well.

The project began to take shape. Our aim was to study the dynamics of a sun-Jupiter-asteroid system and show the existence of instabilities in the movement of an asteroid. It was related to the three-body problem that Poincaré had worked on. Jupiter and the sun were our two large bodies, and our small body was an asteroid. Instead of performing calculations using the actual mass of the sun (1.989×10^{30} kg), we used a common mathematical tool called normalization that allowed us to consider the mass of the sun as $1 - \mu$ and the mass of Jupiter as μ. We weren't interested in Jupiter and the sun as you would see them at the end of a telescope. Our focus was on the behavior of the asteroid. But even then, we were not looking at data collected by astrophysicists. In fact, we were not concerned at all with any specific asteroid that a scientist had identified and tracked. To us, the asteroid was a mathematical object. In some sense, our work seemed very practical. After all, we were talking about objects you could see and measure: planets, asteroids, the sun. And yet it was actually highly theoretical. Between that physical reality and abstract reality stood the mathematician, keeping the physical and the abstract in contact.

Puzzling through the questions raised by the project never felt like work to me. It was more like a quest, or like trying to crack a mystery. We were detectives, hunting for evidence, guided by the laws of mathematics and our instincts. But that analogy isn't quite right either, because it felt more creative than that. We were mathematically constructing the universe at the same time that we were describing it as it was. We came up with the equations to study the motion of the asteroid and calculated how much energy was conserved in the system. We weighted the applicability of different theories, searching for sufficient conditions. I had to teach myself to code in a few different programs to trace possible paths. (Coding can sound daunting, but pretty much anyone with a facility for math and a willingness to read some books can do it.) We consulted dozens of references, looking for ways to refine our method and for work that would support or refute our own. We created Poincaré return maps, which were elaborate and dense with detail. After helping launch me and Joseph, Kaloshin stepped back, answering questions when we had them but letting us take the lead. We were coming up with some fascinating results, but what did they add up to? Slowly, we began to piece them together. We were able to show that an asteroid whose initial orbit is far from the orbit of Mars can be gradually perturbed into one that crosses Mars's orbit. It was a new result—worth publishing.

What excited me was not uncovering something about the physical universe. I didn't share that bias toward the study of space that most people seemed to have. I did not find Mars inherently interesting. What excited me was simply discovering something *new*. No one had ever proven what we had just shown. Maybe the feeling was arrogant. We were not, after all, exactly changing the history of celestial mechanics. But it sparked all the competitive instincts I had, which usually found their outlet only on the football field. And this is what mathematicians did *all the time*. They didn't sit around doing really hard homework, which was my old vague conception of the life of a math professor.

In that moment, I realized that this is what I wanted to do with my life. I wanted to produce new results. I wanted to discover things that no one else had. I wanted to be a mathematician.

I started to carry scraps of paper, which I'd pull out on the team bus or plane or at a restaurant while waiting for a meal to arrive, and jot down ideas for different problems. Little piles covered in mathematical symbols or drawings of shapes grew on my desk and floor. To anyone who saw my room, the scraps looked messy and indecipherable. But to me, they were sketching out something beautiful.

I sometimes thought of an inscription that my dad had written on the inside of a little book he had given me for Christmas during my freshman year in high school, after I had taken that calculus class at the University of Buffalo. The book was called *Q.E.D: Beauty in Mathematical Proof.* It was yellow and thin and small, about the size of my hand. It had not meant too much to me in high school, but I brought it with me to college, and one day that spring I picked it up again. *QED* stands for *quod erat demonstrandum,* or "what had to be proved," and it is traditionally the last line of a proof. The meaning of those words struck me differently now. I heard—I felt—the urgency communicated by those three little letters. What *had* to be proved.

J.C., my dad's inscription read, *to live a happy life, one has to be able to see the beauty that is around us. That sounds easy, but it is surprisingly difficult to do. It requires mental training. Studying mathematics is an ideal form of mental training. Mathematics strips away the dirt of the world to leave the beauty and purity of mathematical reasoning. Enjoy the beauty of reason! Love, Dad.*

||||||||||||||||||||||||||||||||||||||||||||||||||||||||||||

Twelve

Who's Jerry Sandusky?

Football, 2011–2012

That summer, a bunch of guys from the team moved into a house at 808 West College Avenue, on the road that separates campus from downtown State College. Coach Paterno had a rule that big groups of football players couldn't live together, ever since some guys had thrown a keg off the roof of a football house in the 1980s (or so the legend went), but for some reason the coaches relented when my buddies asked. There were nine of them living in the old two-story brick house. After big wins, the whole team—and then some—would gather there. Sometimes around two hundred people would show up. There was furniture in the living room for about a week after they moved in. Then they tossed it and declared the first floor the party floor.

Ty lived there, along with a bunch of our other close friends— Glenn, Adam, Frank, Anthony, Christian, Mark, Drew, and Joe. I didn't live with them, but 808—as we called it—felt like a home for me too. I'd be there on those Saturday nights to celebrate, escaping the jungle on the first floor and taking refuge in the basement with many of our closest friends. Football took up most of our time during the week, but when we got a break, I'd come over to the basement and play video games or hang out with a math book. On warm days, the guys would be grilling out back.

One of the managers on the team mentioned that he played chess, and I challenged him to a game. We started playing twice a week at lunch. There was a purity to it that pleased me. It rewarded precise calculation, but also required intuition, a certain feel for patterns. And it was competitive—for me, *very* competitive. I took winning and losing extremely seriously, knowing that when I lost, I had only my own bad moves to blame.

On the field, I was getting regular playing time. It was my second season as an active player, and I no longer doubted or worried about my place on the team. I had backed up my strong performance against Indiana at the end of the previous season with solid play during spring ball. The coaches were showing confidence in me, and I was splitting games with a senior at right guard. (He played the first and third quarters; I played the second and fourth.) The whole team was playing well. Coming into our bye week—the only weekend that fall when we did not play a game—we were 8–1, 5–0 in the Big Ten. Our only loss was to Alabama, who would go on to win the national championship. We had a shot at making the Rose Bowl.

In late October, we came back against Illinois with three minutes remaining, giving Coach Paterno his 409th career victory. It was the most in the history of Division I football. Still, when the bye week came, I welcomed it. I was beat-up and tired, and I wanted the break.

I planned to spend it hanging out with friends. I also had a stack of books and papers I'd been saving for the weekend off, on a range of subjects: numerical analysis, celestial mechanics, computer science, financial mathematics. I was planning on finishing my undergraduate degree that spring and getting a master's degree the following year, but I was still unsure of what area of mathematics I wanted to focus on. It seemed like a good time to bear down and figure it out.

That Saturday, November 5, I was checking Twitter on my phone when I saw that Penn State had released a statement in response to the arrest of a former Penn State football coach, Jerry Sandusky. I opened the statement, read it, and felt sick: Sandusky had been arrested on charges of sexually abusing eight boys over a fifteen-year period. I couldn't imagine what the victims and their families were going through. At the same time, it never crossed my mind that the news would affect my own life at all. I had never heard Sandusky's name before. I put my phone down and went back to reading math.

A few minutes later, my phone buzzed. It was a group text from one of my teammates. *Who's Jerry Sandusky?*

The old guy who's in the weight room sometimes, another teammate replied.

Later that afternoon, we headed to the football building to work out. That's when we saw the media trucks.

THE STORY WAS ON ESPN, CNN, the front page of the *New York Times*. The focus was not only on Sandusky; the university's athletic director and president were implicated in the scandal for failure to report the complaints of abuse to the authorities. By the end of the weekend they were pushed out. It was reported that one of Sandusky's assaults had taken place in the football building, and that Paterno had learned of it. The scandal began as a story about a man who had done tremendous, unspeakable harm to children, and then it became a story about the university's leadership and the football program. It didn't seem to matter to the public that Sandusky had retired in 1999, or that the events had occurred before any of the players currently on the team had arrived at Penn State. And it *shouldn't* have mattered, because the bad publicity was nothing compared to the sexual abuse of children. Still, the attention—all of it negative, even accusatory— blew us back. We, the players, were implicated and held in part responsible. Penn State football was painted as the problem, and we were suddenly involved.

Reporters camped out in front of the football building, shouting out questions to us as we passed. I kept my head down and walked fast. That Wednesday morning, I was at a lecture for a course on the life of Martin Luther King Jr. when I got a text calling an emergency squad meeting at eleven a.m., less than an hour later. Unsure of what to do, I looked around at the handful of football players who were also in the class. None of them seemed to have seen the message yet. I gathered my things, went over to them, and whispered that we had to go. The professor saw us and stopped the lecture. *What's going on?* he asked.

Emergency meeting, I said. He raised his fist in solidarity.

We went straight to the football building, ignoring the reporters gathered outside the entrance. We took our seats in the squad room, and Paterno made his way to the lectern. Eighty-four years old, he moved slowly. Not holding back his tears, he told us he was gone.

The season continued, surreally. We lost our last two games, and instead of the Rose Bowl, which we had been hoping for, we were invited to the TicketCity Bowl against the University of Houston in Dallas. We lost that too. The locker room was quiet and miserable. Outside of it, there was nothing but noise. We had twenty-three seniors graduating. After Paterno was fired, it was safe to guess that most of his staff would be leaving. Change was coming, and we had no idea what it would look like. Sometimes it was hard for us players not to be defensive about everything. Penn State's Board of Trustees announced that it was conducting an independent investigation, and there were rumors that the football program would be axed. Outside of State College, the rest of the country was convinced that we were tainted by the scandal. The tragedy itself, what Sandusky had done, sometimes got lost in the outrage—by everyone, including us. There were a lot of complicated power dynamics at play, and some of them were out of whack. Like a lot of big institutions, Penn State sometimes struggled with how to balance tradition, accountability, financial realities, old social norms, and its values. But it was hard to take that moment as an opportunity for reflection or to keep perspective. We were kids. The apparent unfairness of the situation quickly started to seem like an injustice.

I wanted to escape it and move on. I faced a dilemma. Since I was graduating that spring and enrolling in a master's program for the following year, I was already thinking about my next steps. Earlier that fall, I had briefly thought about applying to master's and PhD programs at other schools, and I had spent a long time talking to Kaloshin about where I might go. He offered to connect me with the network of mathematicians that he knew at top institutions, but he warned me that it was better to be over-prepared. *There is no learning in a PhD program,* he told me. *There is only reviewing.* What he meant was that I would be better off exposing myself to different subjects in mathematics before moving on, so that I could encounter as much material as possible before I was expected to have it down cold. So I had taken the GRE and applied for a master's in mathematics at Penn State.

That winter, though, given what was happening in State College, I revisited that decision. It seemed like it might be a good idea to look elsewhere. There was even a way for me to continue playing football for another school. Although the NCAA normally requires a transfer to sit out a season in order to retain his or her eligibility, a player can transfer without taking time off if he or she enrolls in a program that his or her original school does not offer. I found two master's programs that suited me well academically and that Penn State did not have: applied mathematics at Northwestern, and computational and mathematical engineering at Stanford. I did not contact those schools' football programs or tell any of my friends or coaches at Penn State, but I decided to apply to those master's programs. I wanted options while I waited to see what would happen at Penn State.

Everything seemed uncertain. In the first week of January, we were told that we had a new coach, Bill O'Brien, the former offensive coordinator for the New England Patriots. We had no idea what to expect. We didn't know anything about him, except that he had Tom Brady's trust—and that a few weeks earlier he had ripped Brady on the sideline during a game. A video of it had gone viral, and we passed the link around the team. *Who on earth is this guy?* we asked one another while we watched him scream and rant at the greatest quarterback in history.

We got our first real glimpse in February, after the Patriots lost the Super Bowl, when O'Brien began a series of predawn training sessions. Our new strength coach, Craig Fitzgerald, was there, wearing a T-shirt and shorts despite the snow. By five a.m., we were running sprints on the practice field, rock music blaring from the speakers. The music was the first sign that Paterno was really gone.

Every Friday morning, while the rest of State College was sleeping, we ran circuits in the dark—shuttle runs, agility drills, speed ladders. We would pair off and wrestle. We'd run until we were gassed. Then came the hard part, the best part: the tug.

At the end of every practice, we would gather in a group, and O'Brien would call out two guys of similar size—usually one from the offense and one from the defense—and have them grab onto the handles on a large blue disk. The goal was to try to pull the other guy ten yards or make him let go of the disk. We would drag each other on the ground, literally dig our heels into the dirt, wrench and yank and twist, do whatever we could to get the other guy's grip to loosen. I loved it. There was something essential about it, competition reduced to its raw essence.

Out there, our pain was a measure of effort, our intensity a reflection of desire. There were days when we just tried to conquer the morning. Those mornings made things seem simple. We let other people worry about Paterno's legacy and the threat of NCAA sanctions. We worried about ourselves and one another.

And after practice was over, seven or eight of us would go to breakfast at the Corner Room, and each get a couple of early-bird specials: $2.95 for two eggs, hash browns, and toast. We'd give the defensive end a hard time for getting waffles instead. It was at breakfast one morning that I realized that I wasn't going to leave.

A FEW WEEKS LATER, Northwestern called to encourage me to change my application from the master's to the PhD program. I told them I had decided to stay at Penn State. Then, in April, after Stanford had accepted me, an administrator from the math department called to ask why I had not registered for their admitted-students weekend, and I told them I wasn't coming. I did not think twice about it. Saying no felt right.

A week later, Stanford called again. The math department was offering me a full academic scholarship, a rare offer for a master's student. It would cover my tuition and room and board, and give me a sizable stipend on top of that. What's more, having an academic scholarship would allow me to walk on to the football team. I would not need to use up one of the program's valuable athletic scholarship spots. In fact, unlike the other student-athletes, I would be getting paid to attend Stanford. I would be able to play for one of the best teams in the country, as part of a program that was celebrated, not marked by scandal. And if I decided not to play football? No problem. I'd graduate with a Stanford degree, without paying a dime and without debt.

Come and visit, the department encouraged. *Pick a weekend. We'll fly you out.* It was too good an offer to refuse. So I went.

I walked around Stanford's campus in a kind of half-daze. I loved Penn State, so much so that I had never minded the miles of monotonous forest and farmland, nor the bitter central Pennsylvania winter. But even someone who was as indifferent to his surroundings as I was couldn't miss how beautiful Stanford was. The palm trees fluttering in the gentle wind, the clear blue sky, the sight of students relaxing on the lawn . . . not to mention the world-class academics. Stanford was the ideal school. The professors and graduate students welcomed me, eager to discuss my work and their own. All weekend, at meals and in offices and while walking around in the warm sunshine, we talked about discrete math, graph theory, machine learning—many of the topics that I was most interested in. My mind was constantly stimulated. I was far less certain about what to do than I had been when I got on the plane to head west.

My mother, naturally, was overjoyed when I told her about the offer. *This is a blessing from God*, she told me. Even my father thought I should go—though, unsurprisingly, for football-related reasons. *Stanford's schemes fit you better*, he said.

I was leaning toward accepting the offer.

Then I flew back to State College. For a few days, I walked around campus thinking I had only a few weeks left there. I still had not told my coaches or friends that I was leaving. They already had enough to worry about. There was some talk that the NCAA was going to try to end the Penn State football program by applying the so-called death penalty—a ban from playing the sport for a period of time, which could effectively kill the program. The public's attention had lessened, but it had not gone away. Guys were worried about keeping their scholarships. The new coaches were under no obligation to keep anyone, let alone to give anyone playing time. The depth chart under Paterno was now meaningless. Everyone had to earn his spot again. That made for an intensely competitive environment, but it also created a team that was closer than any team I had ever been on. We trusted O'Brien because we had no choice. We trusted one another because no one else could quite understand what we were going through.

I decided to stay. I did not know what would happen to the football team. I did not know what would happen to the school or to any of us. Nothing could erase what Sandusky had done. But I loved Penn State, and I loved my teammates. I wasn't going to walk away.

Thirteen

Proof

Math, 2012

On some days, the Sandusky turmoil tested my ability to slip so easily between football and math. It could be hard to find the energy—and the time. On other days, it was a relief to have a refuge. When I was working on a problem, I could tune everything else out. I spent a lot of time that winter and spring exploring different areas of mathematics, searching for something to settle on. But as I did so, I also focused on one thing they all had in common—perhaps the most important thing: proofs. Part of the appeal of math had always been that I could trust that the answers it gave me were true. It wasn't until I started taking more advanced math courses that I wondered what that trust was worth. I had used my intuition, my reason, and the laws that I had learned. For mathematicians, though, that wasn't enough. They demand proof. For a mathematician, that meant *proofs*.

To mathematicians, the word "proof" means something different than it does to most people, even to people whose business is verification. When prosecutors prove their case, for instance, they present evidence to establish the truth of a claim beyond a reasonable doubt, but the doubt does not disappear altogether. There is always the slim chance of an alternative explanation. Or when scientists verify a hypothesis, they show that their assumptions consistently agree with experimental results, but even highly accurate tests can have errors and discrepancies. Not everyone agrees about what "truth" means. A lot of people are suspicious of the very idea of truth, I knew. They might think that truth is culturally constructed, or that there can be contradictory truths, multiple truths, personal truths. They might think that truth is something felt, not proven, or that it can be known only through the worship of God. They might find the subject epistemologically problematic. They just might wonder how much the truth really matters.

Mathematicians have a different standard. What I liked about mathematics was that truth meant something very specific. A proof is an argument about a statement's truth based on reason. This idea became more remarkable the more I thought about it. So many things that I believe in the rest of my life, after all, are subject to disagreement. But mathematical reality, the reality of ideas and reason, is durable. For a statement to be true, it must be true without exception. It doesn't matter whose perspective is adopted, or what moment in history it appears in. It is independent of the person who utters it, or the place where it appears, or who is in power, or what one's religious beliefs are.

That's what makes the idea of a proof so powerful to me. It can be trusted. One of my textbooks, Andrew Browder's *Mathematical Analysis*, began its introduction to the concept of a proof with two models from ancient Greece. (One was the theorem that there are infinitely many prime numbers; the other was the Pythagorean theorem.) The proofs that are written today are not so different from the ones that were written more than two millennia ago. No one believes in the Greek gods of Olympus anymore, but the proofs that Greeks worked out are as true today and useful as they were thousands of years ago. Around 300 BC, the Greek mathematician Euclid wrote a treatise establishing the principles of geometry and the elements of mathematics. Many of his results were previously known, but he created a model for making them conform to a set of axioms, which are simple and self-evident assumptions. He began by listing definitions (a point, an equilateral triangle, a line, a curve, and so on) and five "common notions" (for instance, things that are equal to the same thing are also equal to one another). To that, he added five postulates (any two points can be connected by a straight line, and so on). And with just those definitions, axioms, and postulates, he was able to build theorems. There was a clear and sound system of logic, based on a single, simple idea: if A implies B, and if we know that A is true, then we can conclude that B is true.

As mathematics became more complicated, verifying mathematical statements and showing what could and couldn't be done became even more important. Over time, mathematicians established a system of symbolic notations and formal conventions. It may look forbidding to people outside the field, but it is a language that everyone can speak and that everyone can trust. Gottfried Leibniz, the inventor of calculus (along with Newton), once imagined an actual universal language that people would speak in place of languages such as German or Mandarin. "This language will be the greatest instrument of reason," he wrote. Instead of duels, he continued, "when there are disputes among people, we can simply say, *Let us calculate, without further ado, and see who is right.*" (Leibniz had clearly never sat in the stands during a football game.) Mathematics really does work like that: it resolves disputes once and for all. Every established mathematical statement can be verified with a proof.

Only the techniques differ. One technique works according to the principle of induction. The trick here is to prove a statement for one case and then show that it also holds for the next—and, by inference, the next, and the next, and so on. (It's sort of like a proof by domino effect.) There are ways of proving things directly, building logically from a hypothesis to a conclusion, using only assumptions that have been previously stated or proven. If I want to prove the statement "*If p, then q*," then I would use what I already know about *p*, along with other axioms and theorems I know to be true, to derive *q*. Or, instead of beginning with a hypothesis and deriving some conclusion, I could begin by assuming that what I wanted to prove is actually false, and then show that making that assumption leads to a contradiction. From there, it follows that the original statement is true.

One of my favorite proofs operates this way. I was reading Browder's textbook, *Mathematical Analysis,* one day, exhausted after practice, when I came across a proof that drove home the power of a proof by contradiction. It was worked out by Georg Cantor, who was puzzling through the concept of infinity. For thousands of years, people had thought of infinity almost mystically. In the 1870s, Cantor decided to treat it as a regular mathematical concept, subject to mathematical rules. Cantor thought of an infinite series of natural numbers (1, 2, 3, 4 . . .) and realized that it is countable. If you had an infinite amount of time, you could count every natural number! Then he looked inside the set of natural numbers and saw that it had subsets—for instance, multiples of ten. It turns out that those subsets are also infinite. How do we know? We can assign a number to every subset.

$$1 \leftrightarrow 10$$
$$2 \leftrightarrow 100$$
$$3 \leftrightarrow 1,000$$
$$4 \leftrightarrow 10,000$$

. . .

He was matching the infinite sequence of natural numbers to the infinite subset in a one-to-one correspondence. (Another way of thinking of it is like numbering a list.) Cantor realized that you could even do this with rational numbers (numbers that can be expressed as fractions), even though it might seem like there should be more fractions than there are natural numbers. Both infinities are countable. Then he took on the set of real numbers. Real numbers include not only the rationals but also the irrationals—those numbers that cannot be expressed as the ratio of two integers. When you take a decimal expansion of an irrational number, the numbers go on forever, without any pattern. (The number *pi*, commonly but imprecisely expressed as 3.14, is the most famous irrational number. The square root of 2 is another.) It is impossible to use Cantor's technique of assigning a natural number to every real number. They are not countable. But how could he know it for sure? Cantor was a mathematician, so he needed a proof. He used a proof by contradiction.

The general idea is called the diagonal argument. I followed along the proof with fascination. First, I supposed that the set of decimal numbers is in fact countable—so we would be able to make a numbered list of every decimal expansion without leaving anything out (assuming, of course, that we had infinite time). Imagine an infinite set of base-two decimals following an infinite pattern.

$$1 \leftrightarrow .\mathbf{0}000000000 \ldots$$
$$2 \leftrightarrow .1\mathbf{1}11111111 \ldots$$
$$3 \leftrightarrow .01\mathbf{0}1010101 \ldots$$
$$4 \leftrightarrow .101\mathbf{0}101010 \ldots$$

If you take the first digit from the first line, the second digit from the second line, the third from the third, and so on (the bolded digits above), and turn every bolded zero into a one, and every bolded one into a zero, then you have constructed an entirely new number. We can be certain that that new number is not on our original list because we know, by definition, that the first digit is not the same as the first number on our list, that the second digit is not the same as the second digit of the second number on our list, the third is not the same as the third, and so on. Every single number on the list contains a different digit than our new number. We have a contradiction, so our assumption that the set of decimal numbers is countable is incorrect. It is uncountable. And what that means is that the infinity of real numbers is *bigger* than the infinity of natural numbers.

Every theorem I encountered was accompanied by a proof. I kept coming back to this fact, restating it to myself, contemplating its consequences. We believe that statements are true not because we have strong evidence to back them up, or because we are convinced beyond a reasonable doubt, or because they just feel right and seem to work. We believe they are true because it is impossible for them *not* to be true. They are not subject to biases or cultural influences or changing times or new evidence. No discoveries in the real world will invalidate them. Every symbol, every word in a proof, is precisely defined. Every statement can be traced back all the way to the foundations of mathematics and logic. The situation with the football team could seem out of whack, but in mathematics, there could be no confusion.

OR AT LEAST, that was the idea.

Before practice, while most of the students—and even most of my teammates—were still sleeping, I'd wake up early and work on problems or read. I'd do the same before going to bed. Most of the books I was reading were textbooks, but occasionally I'd come across a book or essay that took a broader perspective. I was always more interested in mathematics than philosophy, more drawn to physics than metaphysics, but some of the deeper questions became inescapable when I started working more seriously with proofs. For instance, if proofs were to be founded on perfectly solid foundations, built only from axioms and statements that have been proven before, where does the chain of logic begin? At the end of the nineteenth century and early twentieth, a German mathematician named David Hilbert had led an effort to formalize proofs so that they existed only on axiomatic grounds. He wanted to build a system statement by statement, one that would make intuition unnecessary. The human element would disappear; there would be no reliance on common sense or innate understanding. We would be left with only well-defined facts. He had given a talk once in which he said, "In mathematics there is no *ignorabimus*"—Latin for "we will not know." He went on, "We must know. We shall know." But as it turned out, that was not true.

A young Austrian mathematician, Kurt Gödel, showed why. He discovered that any axiomatic system that relies on basic arithmetic will be incomplete. There will always be statements about the natural numbers that are true but unprovable within the formal system. To show this, he used a brilliant—and rigorous—proof. Soon after, he wrote a second paper extending the results of the first, showing that a system cannot demonstrate its own consistency. Gödel had been in part inspired by a famous paradox that I had come across as a kid doing puzzles: "This sentence is false." (It's known as the liar's paradox, and it stretches back to ancient Greece.) If the sentence is true, then it is false, because that is the claim that it is making. But if it is false, then it is true, because that is also what it is saying. It can't be both true and false, and yet . . . Gödel created a kind of formal analog to the paradox using self-referential logical propositions to prove that there are true propositions that cannot be proven within arithmetic.

Many people thought that Gödel's results meant the end of objective truth. That horrified Gödel. Like his good friend Einstein, who rejected interpretations of relativity that elevated subjectivity, Gödel absolutely believed in the existence of a mathematical reality. But his work did leave open a role for intuition in mathematics.

I thought about the role of intuition in mathematics a lot. Reason and imagination can influence and reinforce each other. Intuition will not take you very far if you cannot support it with sound theory, but theory has its limits, if it is not constantly questioned and invigorated by intuition. I rely on reason to answer questions and prove statements, but I listen to my intuition—which can lead me to conjectures, speculation, useful analogies, and extrapolation from examples—to know what kinds of questions to ask in the first place. There is a reason that mathematicians are so stubbornly insistent on the idea of "beauty" in mathematics, a word that probably strikes any non-mathematician as very strange in this context (How can a formula be *beautiful?*), but any mathematician immediately understands.

One of the things I read while lying in bed as the sun rose was a short book from 1940 by G. H. Hardy, *A Mathematician's Apology.* He meant "apology" in the old sense—as an argument. The book examined what was useful and useless about mathematics, and why some might make it their life. It made a great impression on me.

"Beauty is the first test: there is no permanent place in the world for ugly mathematics," Hardy wrote. And so it was for me. One of the first things I asked myself when I decided whether or not to pursue a project was not just whether it was "important," but whether it might meet Hardy's criteria—whether it had "a very high degree of unexpectedness, combined with inevitability and economy." I let my intuition guide me down certain paths, often with good results, in search of something beautiful. The challenge in mathematics, I was learning yet again, is not to ignore intuition. It is to have *better* intuition.

Fourteen

||||||||||||||||||||||||||||||||||||||||||||||||||||||||||||||

Sanctions

Football, 2012

We trained that summer unsure of what was going to happen to us. On the morning of July 23, Coach O'Brien came through the weight room as we worked out, telling us that the results of the independent investigation, led by former FBI director Louis Freeh, would be released at ten. Adrenaline surged through me. I looked around at my teammates and could tell they were feeling it too. It was one of my best lifts in a long time.

After showering, we gathered in the lounge to watch the announcement together as a team. Our anticipation turned to stunned silence as soon as the NCAA president, Mark Emmert, began to speak. "Football will never again be placed ahead of educating, nurturing, and protecting young people," Emmert said. A few guys broke the stunned silence by yelling at the TV. "Penn State can focus on the work of rebuilding its athletic culture, not worrying about whether or not it's going to a bowl game," Emmert went on. Sandusky wasn't the real problem, his words implied. *We* were. The football program wasn't cut, but it was effectively wrecked. The university was handed a $60 million fine, and the football team was banned from participating in any bowl games for four years. All of the school's victories from 1998 to 2011 were vacated. The school was put on a five-year probation. We lost ten scholarships a year for the next four years. The team was limited to sixty-five scholarship players on the roster, while other schools were allowed eighty-five. Most devastatingly, anyone would be allowed to transfer without having to sit out a year. We all knew what that meant: coaches from other schools would be circling, trying to vulture our best players. The tightness of our team was its best strength. It was the main reason I had decided not to go to Stanford. But that closeness was about to be threatened in a very big way.

I thought the sanctions would be bad, but I hadn't imagined this—not only the punishment, but the sanctimonious tone. (Within a few years, the NCAA would reduce and remove the team's penalties.) None of us had any fondness for the NCAA, which was profiting off our hard work. None of us in that room thought that football was more important than the safety and well-being of children. We were aware of how serious the situation was, but we couldn't see what destroying the team would accomplish. I knew how important it was for the university and our community to figure out why and how the scandal had happened, and to make sure that nothing like it ever happened again. We needed to think and talk about how power worked in and around the institution, and about how we could make sure that we were more caring, responsive, and responsible. But it hardly seemed like the NCAA's depriving students of scholarships was the way to emphasize the importance of education. Most of those accused of mishandling the revelations of Sandusky's conduct were already gone from Penn State. The severity of the sanctions immediately turned the attention away from the abuse itself and on to us. It had the perverse effect of making the team feel persecuted. What we faced was nothing compared to what those young boys lived with. In that light, our team's fate hardly mattered. But in that moment, we lacked that perspective. It was hard not to feel wronged. I was left to wonder whether our team could survive.

As soon as the press conference was over, we took a quick break. I went over to lunch with a couple of teammates. We were all frustrated. *This shouldn't have anything to do with us,* one guy said. *We had nothing to do with it.* There wasn't much else to say.

When we got back to the football building, we headed to the squad room. O'Brien took the podium, and he spoke to us in a way that made me believe he understood how we felt. He was honest and straightforward, and he didn't try to downplay the extent of the NCAA's penalties or tell us to ignore them. He told us the rules were the rules, and we were free to go. We could stay at Penn State, quit football, and keep our scholarships, or we could go to another team. But he also told us that if we did stay, we would get to play football in front of a hundred thousand people. We'd get to be part of a special group. And we'd get a good education.

Later that day, as I walked to class, it occurred to me that I would not have been at Penn State under the new rules. With a limited number of scholarships, there was no way that the team would be able to take a risk on a two-star recruit like I had been, someone who brought something to the program besides his football skills. Ironically, it was at Penn State where I had discovered how much academics, not just football, mattered to me.

———————

Opposing coaches started calling immediately. They came to State College—not an easy thing to do, given its tiny airport and isolation—and camped out in the parking lot. We had all picked Penn State when we chose where to go to college, but for me, the decision to stay—essentially to recommit—meant something different.

Ty called me a few days later. *You getting these calls?* he asked.

Yeah, I answered.

What are you thinking? You gonna go?

Honestly, I thought seriously about it this spring. Stanford made me an offer—not the football program, the math program. I even flew out there for a visit. But I'm not going.

Ty was silent for a second. *I don't think anyone would blame you if you did.*

I've made my decision, I said. *I'm staying.*

It's hard to know what to do, he said. *We'll be hanging out and someone's phone will ring and he'll go into his room, and you know a coach is calling. What if everyone goes? What if we don't even have enough guys to field a team?*

Could happen, I said.

I just want to play football, he said, miserably.

Me too, I said.

|||||||||||||||||||||||||||||||||||||||||||||||||||||||||

Fifteen

Computers

Math, 2012

Some people imagine mathematics as an escape from the messiness of reality. Some mathematicians do too. And it was true, when I opened a math book or put a blank piece of paper in front of me, I could clear my mind and lose myself in whatever problem was in front of me. I loved the rigor of it, the clarity, the logic. I wasn't after formal purity, though. I didn't want to spend all my time contemplating the properties of prime numbers. I found math beautiful, and that was part of the appeal, but I wanted to explore the gnarled, complex world, not remove myself from it. During my junior year, I was drawn to the types of problems that we couldn't solve perfectly.

That spring of 2012, my last semester before getting my undergraduate degree, I had taken an advanced course on numerical analysis, which is the study of algorithms involving numerical approximations—algorithms to calculate problems that are used when there are too many variables or too many unknowns for the expressions to be solved exactly. Like all branches of analysis, it involves the study of limits. In this case, the goal was generally to show that the algorithms were well-behaved, predictable within some limit, and practically applicable.

The class was taught by a young postdoc named Xiaozhe Hu, who had recently received his PhD from Zhejiang University. He had a thick Chinese accent but spoke English perfectly in an expressive voice. Xiaozhe had a round, unmistakably friendly face, and infectious enthusiasm. I am convinced that it would be impossible not to like him. Perhaps it helped that I was fascinated by what he was teaching.

Numerical analysis has applications in social networks, big data, machine learning. It helps us predict the weather, model financial markets, create battle plans in war, understand the mechanisms behind the spread of disease—situations in which there are too many variables or the calculations are too complex to determine with certainty. We can model the world, but not exactly. Error can be devastating when you're dealing with a hurricane, or deadly virus, or war, in any scenario when lives are on the line. There can be high stakes.

But I didn't have a noble or even practical purpose for wanting to study numerical analysis. I wasn't doing it so that we could predict the paths of hurricanes more effectively, or help stop the spread of disease. I just thought the mathematics was interesting. What appealed to me was the challenge to do it better, and to find methods that were elegant and interesting and opened up new avenues for investigation. We could create new tools that were faster and more efficient.

Computers help, Xiaozhe told the class. *Numerical analysis was transformed by the increasing capacity of computers.* Then he paused and looked at us. *Of course, you are all computers too.*

IT USED TO BE that a computer was a human, not a machine. The word "computer" referred to a person who did arithmetic for a living—for a financial company, maybe, or an astronomy lab. Thousands of (human) computers had jobs in government, business, and the sciences. They not only dealt with account books, but calculated things like the course a ship should take across a sea, the stresses on a dam, the path of a comet. Later, human computers would determine the thrust-to-weight ratio needed to lift a bomber into the air, the trajectory of a rocket's path, and the complex formulas that described nuclear fission. Mechanical aids existed to help them—abacuses, tally sticks, fingers, and, eventually, rudimentary analog machines—but for the most part, calculations were done in the head and on paper.

Then Alan Turing came up with the idea behind the modern computer. A mechanical computer had been sketched out a hundred years earlier by an English mathematician named Charles Babbage, but it had never been built. Turing wasn't thinking of Babbage but of Hilbert, the man who was determined to show that mathematics could be totally formalized and put on logically consistent grounds. Hilbert wanted to find a computational procedure that could show whether a statement is universally valid. Turing had doubts that such an algorithm could exist. As part of his effort to show why, he started modeling mechanical computation. Essentially, he started thinking about a thinking machine. There were already some electronic machines capable of basic computations, but they were not like the ones that Turing imagined. If you wanted one of these primitive computers to do something besides the task it was wired to do, you would have to rewire it. An adding machine was an adding machine; it could not do anything else. Turing had an idea for a "universal" computer that

123

could perform *any* function as long as it was fed the right coded instructions. Turing's universal machine would be programmable.

The story of how such a computer was actually built has been told many times. What amazes me is not the incredible technological advance. It is the powerful thinking of the mathematicians involved, especially Turing and John von Neumann, who helped come up with the architecture for the modern computer. In Turing's case, it was his ability to take an open question and see an open door, to reframe it and see its implications. In von Neumann's case, it was his skillful synthesis, his ability to see the potential in an old idea and take it to a new place. Progress proceeds on both tracks. What they shared—and what inspires me—was the breadth and depth of their knowledge, the quickness of their minds, and their skill in a range of complicated mathematics. When I learned about them, I felt the same way about them as I once had about the Michigan football player Jake Long.

OVER THE COURSE OF THE SEMESTER, Xiaozhe and I became friendly, and then friends. Our relationship was different from the one I had with Kaloshin, who had just left Penn State for the University of Maryland. Xiaozhe had a lot to teach me, but he became more of a collaborator than a mentor. We would share ideas, argue about approaches, and puzzle through problems together.

Most of what we were working on involved algorithms. People often think that the word "algorithm" suggests something daunting or mysterious. An algorithm is basically just a procedure for solving a problem, a list of instructions that begins with an input and ends with an output. It doesn't have to involve a computer. In fact, the term comes from the Latinized name, *algoritmi*, of Muhammad ibn Musa al-Khwarizmi, a scholar based in Baghdad in the ninth century who lived more than a millennium before Mark Zuckerberg learned to code. You don't need to be dealing with numbers to use an algorithm. A recipe for baking a cake is a kind of algorithm. Or you could write an algorithm for eating a sandwich. Maybe it goes like this:

1. Lift sandwich to mouth.
2. Open mouth.
3. Take bite.
4. Chew four times.
5. Swallow.

That should work pretty well, right? But what if you were told:

1. Swallow.
2. Lift sandwich to mouth.
3. Take bite.
4. Chew four times.
5. Open mouth.

Gross. Nobody wants to see that.

The algorithms that mathematicians use in numerical analysis, of course, tend to be a lot more complex. They help create realistic mathematical models of problems that cannot be solved explicitly. The growth of computers has helped make those models much more widespread and reliable. But first, it is worth noting that some things that we ask a computer to do aren't exactly computable. *Computers make approximations too*, Xiaozhe explained in class.

Before I understood how computers work, this might have surprised me. After all, computers can defeat every chess player in the world. They can decode your genome. They can connect Alabama with Azerbaijan with a single click. Surely they can do simple arithmetic! You might be very confused if you asked a computer to evaluate .2 + .1 and, instead of spitting out .3, it gave you the answer .30000000000000004. But in fact, that is exactly the result a computer might give you. It's not a glitch or a bug. The error is a result of the computer's design. When we think of numbers, we usually think of digits that go from 0 to 9—the base-ten system. We already know, however, that that is not the only valid way to represent numeric values. To make certain operations such as multiplication and addition easier, computers store numbers in a binary (or base-two) counting system. The binary system uses just two numerals instead of ten. Every value is represented by groupings of "bits," where each bit represents one of two states (say, 0 and 1—or signal on and signal off). Information encoded in bits can include anything—letters, words, sounds, pictures—as long as it can be represented numerically.

Computers have to make a trade-off between efficiency and exactness. Small errors creep in when computers move between the base-two system of their architecture and the base-ten system we use in everyday life. This is partly because a computer has finite memory. It can't represent a number that goes on forever, so it has to round it or cut it off. Rounding error is easiest to explain in base-ten because that is what we are used to. Consider the decimal representation of $1/3$: 0.333333 . . ., where the 3s go on infinitely. We have only a finite way of storing that number, so we have to stop the 3s somewhere. If $1/3$ is represented by .333333 . . . ., and .333333 . . . is shortened to .3333, then the problems are quickly obvious. If you multiply $1/3$ by 3, you get 1. But if you multiply .3333 by 3, you get .9999—close, but not quite. Using a base-two system, computers run into the same kind of problem. Not all computer problems involve rounding error. But some do.

For instance, computers cannot represent the decimal .1 in binary format with total accuracy. Every time a computer has to round off or truncate a number, a tiny error creeps in. Usually you don't notice it. If the error is in the twenty-seventh digit of a decimal place, it often won't affect your calculations. But in some cases it does, and the rounding errors can accumulate.

One of the first things we had to do in numerical analysis was learn how to account for these kinds of inaccuracies. Otherwise, our algorithms would be sloppy. *Once you know that you have to cut off the number, you need to know at what point to do it,* Xiaozhe explained. Just how much accuracy do you need? How do you account for numbers of different magnitude (really tiny numbers and really huge ones)? Even the attempt to standardize the process, a model based on something called floating-point numbers, can lead to problems. When you subtract nearly equal numbers, for instance, you lose significant digits. Let's say you have two numbers with accuracy to the fourth digit: 1.3333 minus 1.3332. The answer, .0001, however, has only one significant digit. You may say, *.0001 is the right answer!* But we're dealing with numbers that have been approximated in the first place. 1.3333 could have been ⅓, and we have already seen what happens when thirds are rounded. That's a very basic example, but with more complex equations, sometimes programs have trouble finding any significant digits at all. The more I learned about complex functions and different ways to evaluate them, the more I was intrigued. The questions that numerical analysis dealt with were so fundamental, so rich with both theoretical and applied problems. And yet, it also seemed very new. My experience doing mathematical research with Kaloshin had sparked a strong desire to find new solutions, to expand—even in a tiny way—our understanding of the universe. I was starting to see opportunities everywhere.

XIAOZHE HAD COME TO PENN STATE from China to work as a postdoc with a professor named Jinchao Xu, who was one of the most prominent numerical analysis scholars in the world. Jinchao had published several important papers and was almost a legend in the math community in China. He had a vast network of collaborators, working on a wide range of problems at any given time. While I was taking Xiaozhe's class, I was also taking Jinchao's course on numerical linear algebra. It was my first PhD-level class.

For whatever reason, Jinchao took a liking to me. Perhaps it was because I was hard to miss in a math seminar: I was the only undergraduate and one of few non-Asians in the class, and it's fair to say that I was double the size of most of the other students. Jinchao had also heard of me through Xiaozhe. He was impressed enough by my work early in the semester that he asked me to present a seminar on a particular problem in numerical analysis. Xiaozhe came. So did a professor with whom Xiaozhe and Jinchao often worked, Ludmil Zikatanov.

At the end of the seminar, I made a stray comment about expanding ideas of connectivity induced by the Fiedler vector to a more general framework. (A Fiedler vector is a tool in graph theory that helps us see which parts of a problem are most closely related.) *I think you may be on to something,* Ludmil said. I took his comment as an invitation— whether he meant it that way or not. I stopped by Ludmil's office in the mathematics department to ask him a few questions about Fiedler vectors. I went home, read more, and then came back to Ludmil's office the next day. And the next. The following week, I came back again.

Ludmil was a little skeptical of me at first—not because I was a football player, but because he was a mathematician, so he was skeptical by nature. He had never taught me in a class, after all, so he was unfamiliar with my work. But he tolerated my pestering with considerable patience, a patience that turned into interest. It turned out we shared some of the same mathematical instincts.

Ludmil is from Bulgaria. He is a perceptive, creative, and disciplined thinker. When I spoke, he would look at me for a long time, silent and in thought, and then, when satisfied by something I had said, say, *ahhh*. It quickly became one of my favorite sounds.

Meetings at the office eventually became meetings at his house, with espresso and *banitsa*, flaky Bulgarian pastries filled with feta, and CNN on in the background. There were breaks to binge watch British serial crime dramas and for long, relaxed dinners with his wife, Albena, who is a mathematician too, and sometimes with Xiaozhe and a few other mathematicians—some of whom were visiting from faraway places. Mathematics is sometimes portrayed as a solitary endeavor, and it can be. Working on a problem, I would spend hour after hour inside my own head. I was quickly discovering, however, that one of the crucial jobs of a mathematician runs counter to the stereotype: it is to communicate. As satisfying as solving a proof can be, it is even more gratifying to share it with colleagues, who will then put it to use. Each solution becomes the basis for new problems. Not only our papers, but also many of our conversations—sometimes over dinner tables scattered with empty wine glasses, scraps of cheese still on the cutting board—help our understanding of the world. In turn, they help others create better ways of living in it. While mathematicians spend a lot of time alone, our work amounts to nothing if it does not become something that others can build on.

At the end of the spring semester of my junior year, in May 2012, I put on a black cap and gown and proceeded into the basketball arena with the college of science, where I received my undergraduate degree. I won the top prize in mathematics and carried the department's banner as the student marshal—which still ranks among the greatest honors that I have ever received. In some ways, graduation was a different event for me than it was for hundreds of students sitting in the arena around me. For them, it marked a turning point. Most of them would be saying goodbye to friends, leaving Penn State, and embarking on their careers. I wasn't going anywhere but to the football facility.

Still, it was an important moment for me—perhaps not an end, but still a beginning. I was starting to work toward my master's. While I still had a tremendous amount to learn, I was on my way to becoming a mathematician. I could narrow my focus. With sanctions looming, uncertainty unsettled the football team, but I was sure of my path. And I was ready to take the first steps down it.

Sixteen

||||||||||||||||||||||||||||||||||||||||||||||||||||||||||||||||

Who Needs Bowl Games?

Football, 2012

On the Wednesday after the sanctions were announced, July 25, a small group of my teammates went to the practice fields in order to stand with two of our seniors, Michael Mauti and Michael Zordich, as they read a statement affirming our commitment to staying and playing at Penn State. *One man didn't build this program*, they said, *and one man sure as hell cannot tear it down.* I had been scheduled to leave for Chicago for Big Ten football media days, along with two of my teammates, Jordan Hill and Silas Redd, but Coach O'Brien had decided to go alone. It was going to be a circus.

Early the next morning, after a 6:30 a.m. practice, I got a call from OB. I could tell from the tone of his voice that he was about to ask me something he didn't want to ask me, and I was going to have to say yes to whatever it was.

I've changed my mind. If we don't bring any players, we'll look like we're running. Can you come after all?

I wanted to say no. All the questions would be about Paterno and which players were staying and which were going—questions I couldn't answer. I knew why OB wanted me there, of course. I looked calm and collected and had a 4.0 GPA. Without needing to say a word, I was a living rebuttal to the charge that the Penn State football program cared only about football.

I swallowed. *If you need me, Coach, of course I'll be there.*

I rushed home, grabbed my suit, and headed to the airport, where I met Jordan and Mauti, who was taking Silas's place because Silas, our star running back, was thinking seriously about transferring to USC. Mauti was articulate and passionate, and in the past six months he had become one of the team's indisputable leaders. There was no one I respected more. We were in Chicago by lunchtime.

That was the start of my new role as one of the most visible faces of Penn State football. I knew the right things to say, and I said them. *Bowl games? Who needs bowl games? We play in front of a hundred thousand people. I'd rather play a home game at Beaver Stadium than any bowl game.* We talked about our loyalty toward one another, how much we loved our school, how ready we were for the season, how much we wanted to move forward. I meant what I said—but tried to give answers that were far more reassuring than the uncertainty I felt about the future in my heart. It was exhausting.

It only got more exhausting when the season began. The public expected us to lose. It *wanted* us to lose, as if our losing was an appropriate punishment for the terrible things that had happened to those young boys long before we had come. And lose we did. We lost our first two games to teams we should have beaten. It was hard not to be pessimistic early on, when people around the country gloated at every missed field goal, every fumble, every stuffed run.

We tried to stay positive. We would show the world that they were wrong, that we hadn't been killed off, that we were strong. We had a reason to keep playing. We talked a lot about adversity—which, in hindsight, was a mistake. We weren't the real victims in this situation. Our adversity was nothing compared to what those kids faced. Still, that feeling is what motivated us. We were on our own, playing for no one but one another and ourselves.

And it worked. We looked at one another as brothers. We all could have left Penn State. We stayed because we *wanted* to be there. We knew we were probably going to lose a lot of games. When we went on the road, we would be jeered. When we were at home, the pressure on us would be almost overwhelming. Guys could quit or leave, no questions asked. And everyone who decided to stay would be associated with a horrible scandal. Everyone was getting transfer offers. A couple of guys had gotten more than fifty. One of our rivals sent eight staff members to State College to make personal appeals.

Instead of letting them tear us apart, we grew closer. We doubled down on our commitment, and we decided to do it our own way. One of Joe Paterno's old-school rules had been that every player was required to keep his hair short, and beards were forbidden. Only a trim mustache was allowed. So with him gone, we grew our hair long—very long. Ty's hair nearly reached his shoulders. My beard was so massive that it started to grow over my cheeks. Older Paterno supporters would see me on the street and their faces would grow sour. *JoePa is rolling over in his grave*, they'd hiss. Finally, one day O'Brien stopped me in the hallway, looked at the matted mass of hair on my face, shook his head. The deep dimple on his chin twitched as he tried to hide his smile.

I'm getting phone calls about the beards, he said. *Letters. You know I don't care, but I don't need these problems. Do me a favor,* he said. *You mind trimming it back?* I thought about it for a second and nodded. I would do it for OB.

WE FINALLY WON A GAME, and then another, and another—winning seven of nine with one left. Finally, improbably, we reached our final game of the season, against Wisconsin, assured of a winning record. It was senior day, and one of the most memorable and emotional games I've ever played.

It began badly. Four plays into the game, we were down 7–0. We ground out a fifteen-play touchdown drive to tie it, but Wisconsin scored so fast again that it seemed like I had hardly gotten off the field when I had to turn around again and run back to the line. For the rest of the half, our defense settled down—but so did theirs. Running up against their defensive line was like running into a wall. I was directing everything I had, recruiting every fiber of muscle, trying to exploit any leverage, any open angle. But we were getting nowhere.

Then, in the second half, the field started to open up for us. Our quarterback was patient and in control. The line wedged open small holes, and our running back found them. The center, right tackle, and I were working together in perfect harmony. It was like we were talking even as we moved—knowing when to shift, when to share, when to shove forward. The crowd was working itself into a frenzy. It sounded like white noise when I was on the field, and like standing inside a jet engine when I was off. With less than four minutes left, though, Wisconsin had the ball, with 66 yards to go. I was on the sideline. This is the worst part about football: you can affect only one side of the ball. All I could do was watch, wait, and hope we could hold on. Wisconsin was steadily moving toward the end zone. It was fourth and goal from the 4-yard line with 23 seconds left. I was in agony. My nerves were thrumming in tune to the vibrations of the screaming, desperate crowd. The whole season was concentrated into a few long seconds, anxiety and elation felt in their purest form. And of course Wisconsin scored.

In overtime, we had the ball first and brought it close enough to the end zone to go for a field goal. Our kicker's attempt sneaked through the inside of the goalpost. But Wisconsin got their turn with the ball, and they drove right down into field-goal range.

I was sitting with the O-line, holding hands with the right tackle, Mike Farrell. I looked at the ground, unable to watch. There was a long, tense moment of quiet, and then an eruption of joy. I looked up. My teammates were jumping on one another, yanking me off the bench, screaming. That's when I knew Wisconsin's kick had been wide.

You start to think in almost mythic terms when you play football. You start to feel like you can do anything, that you would do anything, that you would risk your life, all to win a game. The man across from you would try to crush you, your body would scream *stop*, and you would go on. We called ourselves warriors and we called one another brothers. It was a brutal, violent, pointless, painful game, and I loved it.

You can believe that we celebrated that night. It was the best win of my life.

Soon after the end of the season, I met with my position coach to discuss my performance that year and look ahead to the next. *I think you should stay another year,* the O-line coach told me. *If I were you, I wouldn't declare for the draft this year. You'd benefit from another year of starting in college. The NFL can wait.*

I looked at him, startled. *The NFL?*

That wasn't the first time someone had said that I had a shot at the

NFL. My dad had been whispering it in my ear since high school—but he was, of course, my dad.

I did know that I was a good football player. I was an anchor on an offensive line that had turned out to be part of one of the best offenses in the Big Ten—despite predictions that we would be dismal. In November, I had been selected to the All–Big Ten first team—an honor that I had dreamed of as a kid, back when I was desperate to be a Michigan Wolverine. I was a first-team Academic All-American as well. My technique had vastly improved since I'd arrived at Penn State, and the new strength and conditioning program under Craig Fitzgerald had transformed my body. I'd gone up against guys who were heading to the League and held my own—or gotten the better of them. Still, until that conversation with my position coach, I had never truly believed I had a shot at the pros. I went home elated. There was no question of declaring early—even if my coach would have encouraged it, I had no desire to leave Penn State a single day earlier than I had to. But I left the football building with a new sense of purpose, a mission.

When I got home, I opened my laptop to continue working on the paper I had begun with Ludmil. We were planning on submitting to a top linear algebra journal later that year. Suddenly, surprisingly, I had a strange feeling: I felt torn.

‖‖‖

Seventeen

Graph Theory

Math, 2013

Until that moment when I started thinking about playing in the NFL, there had been only one year separating me from pursuing my PhD, the next step to becoming a mathematics professor. I was even teaching for the first time that spring, Trigonometry and Analytic Geometry, the last math class that many of the students would take. There was not a single doubt in my mind that I wanted to become a mathematics professor. I wanted to solve significant problems. I wanted to make my mark on the field. I had come to love mathematical research in a way that I could never have imagined. It's a cliché to call something a "passion," but that's what it was. There was something almost romantic about it. I could obsess over a problem for days, for weeks, thinking of nothing else, the way someone might obsess over a girl. But no girl I had ever met brought me the singular sense of engagement that I got from proving something difficult. That sense of satisfaction was so sweet, so deep, so full of wonder. What was more, I knew that I was far more talented and promising as a mathematician than as a football player. There was a chance that some team would draft me, but my chances of becoming a starter or a star in the league, I believed, were small. My potential in mathematics was much greater. I knew I had an unusual gift. That did not mean I would become a great mathematician, of course, but it gave me a reason and the motivation to work hard.

I leaned back in my chair and told myself to slow down. I told myself that if I played in the NFL, I would still be able to do math afterward—but I knew the truth was that I did not have that much time to get started on my career as a mathematician. Football is a young man's game, but so is math. There is a reason that the major award for achievement in mathematics, the Fields Medal, is awarded only to mathematicians under the age of forty.

So I decided to keep doing both, even if I made it to the NFL. At Penn State, I had prided myself on never picking between math and football. Why should the NFL be different? There was an off-season. There were bye weeks. I already had collaborators. Who said I needed to be enrolled in a program in order to do mathematical research, or to study and learn? As far as I knew, there had never been an NFL player who was also a mathematician. Fine. I would change that.

But I was getting ahead of myself. *Way* ahead of myself. Thousands of top college players are convinced that they'll make the NFL. Only a fraction are drafted. Only a fraction of *those* make an active roster. And the winnowing goes from there. It was going to take a tremendous amount of work even just to make the first step. But I woke up every morning with that dream. It got me out of bed at five in the morning, to lift. Over spring break, I stayed in State College and worked out with Fitz at 4:30 a.m.—his normal workout time. We'd meet in the empty gym. I'd look at my reflection in the window, turned into a mirror by the darkness outside, and give myself a silent challenge. Then I'd put 500 pounds on my back.

I'd get home, collapse into bed, and then get back to math.

THE PROJECT that I was working on with Ludmil was never far from my mind. He and I would meet every few days to discuss our ideas or papers that were potential references. We'd write each other epic-length texts in the language of math. As our thoughts became more coherent, we'd type them up into a document on a shared server. The project took shape. It was about networks and graphs—not, however, the kind of graph you may be thinking of (an x-axis, a y-axis, and a line). In graph theory, a "graph" refers to a set of discrete objects—things that are individually separate—and the relationships between them.

People see graphs all the time, usually without realizing it. You can think of a simple road map as a kind of graph. On this graph, each city is a vertex (also known as a node or point). The cities, or vertices, are connected by roads, which we can call the edges. The graph could be extended to contain more information—for instance, the distance between cities. In the language of mathematics, this would be known as a "weight." A road that goes over the mountain is harder to construct than a road through a desert. Such an edge is said to be "expensive" (in math as well as life).

As it happens, graph theory began as a problem on a map—and as a famous puzzle, precisely the kind of puzzle I loved so much as a kid. In the eighteenth century, Königsberg, Prussia, was divided by a river, and in the middle of the river were two large islands that were connected to each other and to the mainland by seven bridges. It was a popular game to try to figure out how to traverse the four landmasses (the two islands and two riverbanks) by crossing each bridge only once. No one had ever been able to do it. When the great mathematician Leonhard Euler visited the city, he decided to prove why it was impossible. No good technique existed, so he decided to invent one.

He did it the way a good puzzler would. Euler's first insight was to turn the problem into an abstraction. He realized that most of the information he had was extraneous and distracting, so he focused on the elements that mattered. Basically, he wiped the names and labels off the map and turned it into a collection of lines connected by vertices, or edges connected by nodes. It didn't matter where the nodes were— whether they were to the north or south, or right or left, or how far apart they were from each other. Nor did it matter whether the edges were curved or straight. Edges could have lengths (another way of saying they can be weighted), but Euler did not care about the coordinates. Then he realized that, with the exception of the starting and finishing points, when you entered a landmass by a bridge, you had to leave it by a bridge. In order to enter and leave that landmass crossing each bridge only once, therefore, there had to be an even number of bridges touching it. Otherwise, you'd have to cross a bridge twice, violating the rules, or be stuck on the landmass forever. However, each of the four landmasses (the two banks and the two islands) was touched by an odd number of bridges. The challenge couldn't be done.

Graph theory grew out of generalizing that seven-bridges problem. The possibility of walking through a graph, tracing each edge only once, depends on the number of edges touching the nodes. (In the seven-bridges problem, the nodes were the landmasses and the bridges were the edges.) Euler showed that the graph had to be connected and have either zero or two nodes with an odd number of edges touching them. That "walk" is now known as an Eulerian trail. If a circuit (or a trail that begins and ends at the same vertex) exists that traverses every edge exactly once, then it is known as an Eulerian circuit.

The bridge problem can be represented by the picture below. Each landmass is a node, and each path crossing a bridge is an edge:

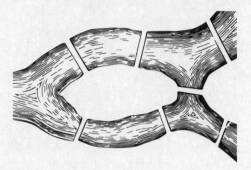

Now that can be abstracted as a graph:

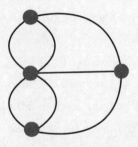

It turns out that Eulerian paths and circuits are quite useful in many fields—even to reconstruct sequences of DNA—and graph theory is widely applied. Even Facebook is a graph, where the users are the discrete objects and friendship is the binary relation. (Facebook defines it this way as well. It has called its network of relationships a "social graph.")

I sometimes get a little defensive, though, when someone points to the real-world applicability of something like graph theory as a reason for why it "matters." The practical uses of math matter, but these are only signs of math's significance, not its justification. Graph theory opened up new areas of inquiry in math. It led to a greater ability to represent the relationships of objects, and a greater understanding of how those relationships work.

THAT WAS WHAT LUDMIL and I were trying to do. I knew that our work could have widespread applicability, but that is not what motivated me to do it. I was interested in how it connected to and deepened and complicated other mathematical ideas.

Sometimes a graph is relatively simple. Imagine a group of people, each with a different task—for instance, a football team. Some of those tasks depend on one another, but not all. The cornerbacks and the safeties need to coordinate their coverage; wide receivers and the quarterback need to be in sync. But the quarterback and the safeties probably don't need to be in constant communication. If you were in charge of the team, you'd probably want to organize the group by dividing it into smaller groups of people who needed to work together, so that every task could be completed as efficiently as possible. (And in fact, football teams do: that's why there are offensive and defensive coordinators, offensive line coaches, special teams meetings, and so on.) If you were to graph this, the tasks would be the vertices, and the edges would be the relationship between the tasks. Grouping the vertices (tasks) that have the most edges (relationships) between them, with only a small number of edges connecting the different subgroups, is called graph partitioning or clustering.

That's a simple graph. Some graphs are far more complex. Let's say you have a computer that is trying to run an algorithm that involves an enormous number of tasks—so many tasks that it would take an unreasonable amount of time for one computer to do them. What if we could break that algorithm into parts that could be run simultaneously on different processors, and then recombined at the end? That's essentially what parallel computing can do. We used a method that let us see what parts of a problem are most closely related, so that it could be broken down efficiently, run on different processors, and recombined at the end, so that a difficult problem would be made less expensive.

If we wanted to divide the group into only two groups, with each section containing the most closely related things and with the fewest connections possible between the two parts, we would make a graph bisection. It turns out that the optimal way to cut the graph would take so long to figure out that it is realistically infeasible. There are, however, several ways to approximate the best bisection in reasonable time using the properties of the graph. An approach that turns out to work pretty well uses something called the spectral bisection algorithms. (Spectral graph theory refers to the study of certain properties of a graph, properties that are related to the matrix representation of a graph.) The method was well known. But there were certain conditions in which the connectedness of graphs was less well understood.

Ludmil and I believed that we could do better. We were convinced that we could refine the existing results, and furthermore that we could say something about the connectedness of the subgraphs in *all* cases. We asked, What if we assume the *worst* possible conditions? I had a hunch that we could show that the two subgraphs would still be connected. No one had ever been able to do it. That sparked my competitiveness. I wanted us to be the first.

I WORKED ON THAT SPECTRAL bisection paper with extra urgency that spring, 2013, because I was determined to include it in my master's thesis, due in May, along with two other papers involving graph theory. (One of the papers refined a technique for the pricing of a financial instrument, barrier options, that was better suited to parallel computing than other methods that were often used. Another was the numerical analysis project that I had been working on with Jinchao and Xiaozhe.) The spectral bisection project was the most ambitious, and it was the one I was most excited about. But it was also in danger of not being finished in time.

That's one of the frustrating things about math. Solving a problem or proving a theorem is not always a matter of discipline and effort. Sometimes, you can work diligently for months and then have to walk away. Some problems haven't been solved yet for good reason. I couldn't force it, but I did my best to try. I managed to finish the proofs for the major theorem and write up the results just before the deadline.

I was proud of my thesis (though not quite as proud, if I was being honest, as I was of being voted to the All–Big Ten first team). I knew the work was of high enough quality to be published, and I knew that it was rare for a master's thesis to include three publishable papers. For once, I didn't focus on the next goal. I just let myself enjoy my accomplishment.

I was talking to my mom on the phone when she asked about my master's graduation ceremony. *When is it?*

I don't know, I answered.

What do you mean, you don't know? she demanded.

I don't think I'm walking, I said. I had already graduated once, I figured, and I wasn't so big on ceremony anyway. I had other things to do. *I already walked once.*

You're not walking?! Mom squawked. *That was your undergraduate ceremony, and this is your graduate one. You are walking, and I am coming.* My mom is not a woman easily denied. I walked, and sure enough, she was there watching, her smile wide.

FOR A COUPLE OF MONTHS, I relaxed. I spent time with my friends. I went fishing at a friend's lake house. I watched horror movies, did yoga, and worked with a mixed martial arts trainer. Then, in July, I started preparing the spectral bisection paper for submission to an academic journal. As I sat in my room, reading through it, a feeling of unease crept over me. The unease turned to panic, and then the panic to despair. Something wasn't right in the proof of the most important theorem. I could see a complication. There was a mistake.

Frantic, I texted Ludmil and asked him to look at the proof. *Maybe I'm missing something,* I wrote hopefully. But I knew the truth. There was an error.

It wasn't a small error either. It wasn't a typo or something that could easily be fixed. The proof did not work as it was written.

I thought of my thesis sitting on the shelf at the library, permanently a part of the Penn State archives. That little error was like a spot of rot, infecting the whole thing. Mathematicians make mistakes all the time, of course. That is a large part of the reason there is a rigorous peer-review process for submission to academic journals. But it is every mathematician's nightmare to make a mistake that slips by the reviewers and into the record. Even if no one else noticed, even if no one else cared, I knew it was there.

Waves of nausea washed over me. Even worse, I did not know how to fix it. I did not even know if it *could* be fixed. It was entirely possible that I had spent months immersed in a problem that I could not solve. I *thought* there had to be a way of approaching it, but now I realized it was only a hunch. I had a long way to go.

Eighteen

Senior Season

Football, 2013

I tried to shove the problem out of my mind a few days later, when training camp began. I had been looking forward to the start of football, determined to enjoy my final season at Penn State, but as camp started, I felt a new kind of pressure. We were still undercut by NCAA sanctions, still barred from bowl games, but we had finished the end of the previous season with a lot of momentum, and I wanted to help keep that going. I always felt nerves before the start of training camp, but there was something else weighing on me this season. In every game I played, I would be essentially auditioning for the NFL. Money was on the line now.

Everything was changing fast. Our season opener, at the end of August, against Syracuse, was played at MetLife Stadium, the home of the Giants and the Jets. When we got to the hotel, I felt a sudden panic when I realized that I had been given a different room assignment than

Ty. I searched for the person in charge of organizing the trip. *Something's wrong*, I said, handing him my room key. *Ty and I are in separate rooms.*

Seniors get their own rooms this time, he said. *No roommate. Special treat. Enjoy it.*

No, I persisted. *I have to room with Ty. I always room with Ty.* He looked at me oddly but made the switch. How could I explain that I wasn't quite ready for things to change?

The growing attention on me brought another kind of pressure. I was held up as one of the faces of Penn State—literally. The school began a campaign called Faces of Penn State to counter the negative image of the school in the national media, and I was one of the students they picked to feature. When I walked down Allen Street in the center of town, I passed my portrait on a banner. Not only the local newspapers but places such as *USA Today* and *Sports Illustrated* were calling me a role model. I was asked to give the keynote speech at the Big Ten kickoff luncheon. I went to what felt like an endless stream of events, gave interviews, shook hand after hand, smiled and small-talked—all the while trying to suppress the anxiety that sometimes swelled up in those kinds of social situations.

I was supposed to "redefine football culture." All this made me feel a little uncomfortable. Redefining football culture seemed like an awful lot to ask of one person. Even if I had been able to, I might not have wanted to. Football culture—maybe not as it was imagined by the public, but as I experienced it—was also *my* culture. My teammates were most of my closest friends. I didn't always feel or act like a great role model myself. I did math and I played football not because I wanted to be a good, well-rounded person, but because I wanted to do math and play football. Sometimes I got tired of it—or just wanted to fool around with my friends—and I'd have some fun with it. Once, an ESPN producer interested in my math pursuits asked me if anyone else on the team had talents outside football. *You should talk to Ty Howle*, I said. *He's from North Carolina, a rural town, and he's an expert in jam. You should ask him about making jam.* Once, I told a reporter to ask Ty about the scar on his shoulder, which he'd gotten from surgery after a football injury. *It's a shark bite*, I explained. *Ask him about getting attacked by a shark.*

At the same time, I accepted the attention and even did what I could to cultivate it. I knew that it would help Penn State, and I was loyal to my school. It had changed my life for the better in so many ways. The spotlight could be exhausting, but I didn't question it or resist it. There was no point in that. The situation was what it was, and I understood it. I recognized the opportunity I had to dispel the idea that an athlete couldn't be a student and a student couldn't be an athlete, the idea that the phrase "student-athlete" was a joke. So I came into my senior season with an extra sense of urgency. I wanted—I needed—to play my best.

I made a conscious decision to commit myself. I became more vocal in the locker room. I tried to set an example in the weight room. I had a sense of purpose.

Then, in the first game of the season, I got hurt.

I WAS USED TO PAIN. Even before I was injured, I dealt with it almost every day. All football players do—especially ones at certain positions. Being an offensive lineman is like being in dozens of small car crashes, only you're not the driver, you're the car. (Being a running back is like getting in dozens of multi-car pileups.) On a goal-line push, the quarterback would try to ride my back into the end zone—and I would end up with more than a metric ton of weight on top of me. I know guys who have played with a torn labrum or rotator cuff. If someone breaks his wrist, there's a good chance he'll be on the field wearing a soft cast the following week. It's part of the game. You bear it, knowing that you won't fully heal until after the season, when you can finally rest.

I'd been lucky, very lucky. Yes, my knees clicked like unoiled gears when I bent them, and a few of my fingers were becoming crooked— but I'd never really been hurt. During that game, though, I felt something yank in my hip. At first I just ignored it, like I ignored everything else that might bother me, but the pain got worse. I'd tell my body to run, and my hip would waylay the signal from my brain to my legs. I could barely practice. Only during games, when nature's greatest drug—adrenaline—kicked in, could I move effectively, and even then, my movement was hampered. I started getting shots of Toradol—a nonsteroidal anti-inflammatory often used after surgery to deaden the body—before games. That was not unusual; I knew guys who took Toradol before every single game for years. I didn't have a second thought about whether it was the right thing to do. After a few minutes, a vague feeling would replace the pulse of pain in my hip. To me, Toradol was a miracle drug.

Still, no drug or level of will can restore a broken body to a hundred percent. It's a strange feeling to expect your body to do something—to know you've done it a thousand times—and have it revolt. Even being a few milliseconds slower, a tiny bit weaker, makes a difference. I'd pull left, feel my leverage was a little off, grit my teeth, and try to will myself to make up the difference. Waves of frustration would wash over me after every play. The stakes felt so high. My coaches and teammates were counting on me. Millions of people were checking the game scores online. NFL scouts were watching. There was my not inconsiderable pride. Even if the injury had been worse, there was no way I wasn't going to play.

My hip didn't properly heal, but I did start to play better as the season wore on. I figured out how to compensate. My instincts were strong. By then I'd been playing football for a decade. I didn't try to do anything fancy. It was set, spring, see, react. Thinking never helps anyway. My intuition would read the situation, registering countless changing variables and concluding how to adjust. I'd make calculations, separate the signal from the noise, see patterns in the chaos—but not like I did as a mathematician. The game moved too fast. I didn't think. I didn't feel. I just acted. I paid for it hours after the games, when the adrenaline and the drugs wore off, but it was worth it.

The pain was small, anyway, compared to the strange mixture of emotions that overwhelmed me as the season went on. Penn State might have been a pariah in the eyes of many people around the country, but to me it was home. Colleges are the great engine of nostalgia for a lot of people, and that can be kind of strange—especially when you're a twenty-year-old kid who's burdened with those memories and still-held dreams every time you try to make a block. But I felt it too, especially in those final few weeks of the season. My jersey meant something to me. The singing in the stands, the ringing of the bell after games. My best friend lined up to the left of me, another lined up to the right. The blueberry muffins at the local diner, sliced and grilled. The grass at Beaver Stadium, which is, I contend, the softest, sturdiest, best grass in the world. I wanted to play in the NFL, but I did not want to leave Penn State.

We came into the end of our season with a 6–5 record—better than anyone could have hoped for a team under the kinds of sanctions we were under, but not quite riding the wave of euphoria that we had experienced at the end of the previous season. The shock of the sanctions had worn off, and it was no longer quite us against the world. Still, we felt that we had a lot to prove—including our independence.

We were playing for one another and playing for ourselves. Penn State football was built on tradition, but we were making our own way now.

OUR FINAL GAME OF THE SEASON, my last game playing for Penn State, was at Wisconsin, which was 9–2 and in contention for a spot in one of the major bowls. We were 24-point underdogs. It was the best thing that could have happened to us. All week long, we walked around happily angry to be so disrespected.

Motivation is impossible to quantify and almost as impossible to talk about precisely. I dislike it when commentators on television say that one team "wanted it more" than the other team. That's absurd. Everyone who makes it to the college or pro level is rabidly competitive. Everyone wants to win. Within the population of elite competitors, I would bet that there is a weak correlation between the desire to win and winning.

That said, we really did want it more. We could have been playing the New England Patriots. We were *not* going to lose.

On our opening drive, our freshman quarterback, Christian Hackenberg, rolled out right and threw a 68-yard touchdown pass. Still on my feet, my job done, I had the perfect view of the receiver running into the end zone. At the half, the score was tied, 14–14. Then the game broke open. We ended up on top, 31–24.

In the locker room afterward, one of my closest friends, Glenn, grabbed a pair of scissors and cut Ty's long hair. Then he chopped off the hair of one of our buddies, Adam Gress. It was over. We were done.

I took my jersey and my helmet from my locker to give to my mom. I never kept anything football-related for myself—no trophies, no newspaper articles, no game balls, no jerseys. Everything went straight to my mom, who would have killed me if I threw a single thing away instead of giving it to her. She still pestered me about when I was going to quit football, and she still wasn't quite sure what exactly I did. When people asked her what position I played, she'd say, *He's next to the guy who gives the football to the quarterback*. But she became a huge Penn State football supporter, and she had driven to State College for every home game.

It was an emotional time for all of us. For most of the other seniors, that was the last football game they would ever play. It is hard to overstate what that meant. Until that moment, football had structured their lives. It had defined them. And now it was over. A couple of guys were going to keep training, hoping to make it into the League. A couple would make it onto rosters as free agents. But the others would be looking for jobs. Even O'Brien was leaving Penn State, heading to the NFL to coach the Houston Texans.

I wasn't through with football yet, of course. But there was still a sense of ending for me in that locker room too. Whatever happened next was going to be very different from what I had experienced at Penn State. That was as it should be. I am not a particularly sentimental guy. Still, it was impossible not to feel sad and a little strange. I sat in the stall of my locker room for an extra few minutes, not wanting to say goodbye.

||||||||||||||||||||||||||||||||||||||||||||||||||||||||||||

Nineteen

Spectral Bisection

Math, 2013

The mistake in my thesis haunted me that fall, my final season of football at Penn State. I thought about it while walking to practice, while eating lunch, while on the bus. I wanted to fix it, and to redeem myself by publishing the paper in the top journal for linear algebra. I would be hanging out when a little lead, a hint of an idea, would nag at the corner of my mind, and I'd stop whatever I was doing or saying and try to think. *Hey Ursch,* my friends would say, looking concerned. *You okay?*

Thinking, I'd reply.

Eventually they stopped asking.

I'd lock myself in an empty classroom overnight, filling the chalkboards with statements that seemed promising but went, finally, nowhere. Every once in a while, one of my teammates or friends would check on me, bringing me something to eat. They'd look at the strange symbols on the board and ask me worried questions; I could tell they were checking to make sure I hadn't lost my mind. I'm not sure they left convinced I hadn't. I went through reams of printer paper, scribbling notes. Sometimes I'd be in a sleepless, euphoric state. I wouldn't leave even to eat. Instead, I would order food to be delivered to whatever classroom I was in. Some days, the only person I would speak to outside of practice was the takeout delivery guy. I wasn't willing to give up on the problem. I had sensed that I was close to a breakthrough, but I couldn't find the way forward. The weeks passed, and then months.

One Friday in late October, the night before we played Ohio State, I lay on my hotel bed in Columbus, working on the proof—or pretending to. In fact, I was just sunk in my own despair.

Ty was sprawled on the other bed, watching TV. He looked over at me. *Hey John,* he said. *You all right?*

Just stuck on this problem, I said. *I'm close, I can feel it, but I can't get it.*

I'm going to give you some advice, he said in a fake-solemn voice. *Step back and attack it from a different direction.*

Thanks for the advice, I said semi-sarcastically. *So wise.*

In mid-November, the night before taking on Purdue, we were back at the hotel in State College. Ty was on one bed, watching the Food Network, and I was, as usual, on the other, with a pen in my hand and a crowd of ideas in my head. Crumpled papers with notes for the half-finished theorem were scattered around me. I stared at one of them, chewing my pen cap, and then started to write. Then I slammed the pen down.

Ty turned toward me, startled. *You done for the night?* he asked.

No, I proved it! I proved this theorem! I nearly shouted. I couldn't contain myself.

That's so great! Can you explain it to me?

Ty was always asking me to explain things to him. I always tried, and he always stopped me halfway through. This time was no different. It didn't matter. We were both so fired up. He seemed as happy for me as I was—and I was as happy as I had ever been in my life.

Are you going to put my name on there? Ty asked.

I started laughing.

What about the acknowledgments? he persisted.

What did you do? I responded, still laughing.

I'm the reason you solved it! Ty said. *It was my advice! I got you going in the right direction!*

No way, man, I said, and smiled.

The paper, *Spectral Bisection of Graphs and Connectedness,* was published in a top journal for linear algebra in 2014. A couple of years later, the American Mathematical Society, the main association of mathematicians, ran a story about me in its magazine, *Notices.* They printed the theorem and had the kindness to name it: *the Urschel-Zikatanov theorem.* "Let G be a finite connected undirected weighted graph without self-loops. For an eigenfunction f of the Laplacian of the smallest possible eigenvalue, the sets where f is nonnegative and negative are both connected."

I ALSO TAUGHT MY SECOND CLASS, Vector Calculus, that fall. There was only one section, which meant I was in charge. I got to not only lecture and grade and hold office hours, but also assign the homework and write the exams. It gave me a chance to think about how I had often been taught—the emphasis on passive learning and regurgitation, the weariness that seemed to infect a lot of the students and teachers. I thought about doing puzzles alone as a kid, and the mentors I had found at Penn State. I wanted to talk to my students in a way that would make them want to listen.

Most of the students in the class were engineering students, and this would be the last calculus class that they would take. I wanted them to leave with a deep understanding of how the math worked—not just a list of formulas to memorize and apply. Every math professor, of course, will say that she or he wants that. But how do you make it happen?

For starters, I wanted the lectures not to be lectures. The class was small enough that I knew all the students by their names, and I used them. What did they think? Could they see another approach? I encouraged them to challenge me. If they saw a different way of doing something, I wanted to know. If the technique that I was proposing seemed confusing or wrong, I wanted them to tell me. I let them know that I didn't always have the answers—which was true.

The real chance for them to learn, I figured, would come when I wasn't there telling them what to do. They were going to have to build their own tools and figure out when to use them. After all, when they were working on a project, no one was going to tell them, *And now use the mean-value theorem to calculate.* . . . So instead of using homework in order to review and check what they'd learn in class, I assigned a few questions that covered the material explicitly, along with other problems that were not only harder but may have seemed off topic. I wanted my students not only to be able to remember what we had done in class, but also to think creatively and to figure out how the more challenging problems were connected to what we had covered. My exams were like that too—only even more so. They weren't just an opportunity to evaluate progress. They were where the students were supposed to do their deepest thinking. My exams were hard—really hard. They demanded that students think creatively and figure out how to apply what they had encountered in textbooks and class. I graded very easily, though. What I wanted to see was not a student's capacity for memorization but evidence of some effort and imagination.

I like to think it worked, at least for most of them. I could see clearly not only how they took bigger risks in their work, but also how they grounded those risks in solid, logical reasoning. The quieter ones became a little bolder about speaking up and questioning me. The ones who thought they knew everything became a little more skeptical of their own genius. A couple of them came to office hours with problems of their own they were interested in exploring. I was having fun. I wanted to be there—and as simple as it sounds, I think that helped too. Passion can be contagious.

The media heard that I was teaching a math course and loved the story. I got interview requests; occasionally a reporter would ask to sit in on a class. The stories were written in the breathless tone that at once made me a little uncomfortable and seemed a little silly. *Football player is smart!* The students themselves, though, were more like me: they did not seem to think it was a big deal. A few of them would show up a few minutes before class on Monday morning and give me a hard time about whatever had happened in the game that weekend. Others signed up for the class with no idea that the teacher was a member of the football team. *I thought you were kind of big to be a math teacher,* one student told me when he realized who I was.

WHO WAS I? I was thinking about transitions and transformations that fall—how to navigate them, how to preserve what was important.

In math, transformations are sometimes represented by matrices, those rectangular arrays of numbers or symbols that I first encountered as a kid in the University of Buffalo library with my dad. Matrices give us a way of changing coordinates. They show how something rotates or stretches or shrinks or shifts. The other thing about matrices, the really beautiful thing, is that a complicated matrix has an essential property inherent to it. It is the root of its stability, what's left when the complexity is reduced. It is called an eigenspace.

Using it requires a bit of advanced math—which makes sense, since it deals with complexity. Conceptually, though, it sounds a lot more forbidding than it is. The idea behind an eigenspace, in fact, is simplicity itself. The word comes from the German *eigen,* which means "own." Think of it as characteristic, or root, or essential. It is the element that doesn't change when everything else is in the midst of transformation.

One way to imagine an eigenspace without getting into the operations of matrices is to think of it as an axis. Every vector—which, for our purposes, we can think of as a collection of numbers in which the order matters—along that axis is known as an eigenvector. They are the vectors along which the transformation does not change direction. Every eigenvector has a corresponding eigenvalue, which tells us how much something is stretched or shrunk.

I used eigenspaces all the time in my work. Eigenvalues, eigenvectors, and eigenspaces allowed me to break down complex systems. They helped me untangle variables and let me see how the linear transformations worked on different parts. They helped me determine whether a system of linear equations has a solution or not, or let me reduce noise in a set of data, or allowed me to simplify things that seemed complicated.

Most people never encounter eigenspaces. Most people don't take linear algebra. And, frankly, that's fine—most people don't *need* linear algebra, not even to be sound thinkers with a good grasp of mathematical reasoning. But everyone undergoes transitions and transformations. Everyone changes coordinates. I came to see eigenspaces as a way of orienting oneself. They were a way of finding what was stable in the midst of change.

Twenty

|||||||||||||||||||||||||||||||||||||||||||||||||||||||

Combine

Football, 2014

After the 2013 season ended and Ludmil and I finished the paper, I turned my focus totally toward training. I was determined to make the NFL—and when I am determined to do something, I tackle it with full intensity.

Naturally, I also assessed my chances mathematically.

If I was going to make an NFL roster, I knew that my chances would be vastly better if I came in as a drafted player instead of signing with a team as an undrafted free agent. In the press, you hear a lot of inspiring Cinderella stories about guys who are overlooked in the draft and go on to have fantastic careers, but those stories are extremely rare. At training camp every August, there are 2,880 men vying for 1,696 roster spots in the NFL. Of those 1,696 spots, I knew that around fourteen hundred would be taken by players who had been previously active on

NFL rosters or on injured reserve. Another two hundred or so would go to rookie draft picks. That left around a hundred spots for undrafted rookie free agents—spread across all thirty-two teams. (In fact, that year, 2014, only sixty-three undrafted rookie free agents would make the final roster cuts for the first regular-season game.) So my first goal was trying to convince a team to draft me.

I had already done the most important part. Teams would focus on what I had shown during my college career. But that wasn't all. I passed the first hurdle when I received an invite to the NFL Scouting Combine. There I'd be measured, tested, examined, and interviewed along with hundreds of other draft hopefuls. For some teams, I knew, what happened at the combine wouldn't matter much, while others valued it more. There are a few instances every year in which someone vaults up the draft board because they show some unbelievable feat of speed or strength at the combine. Coaches absolutely love to talk about "potential." Maybe it flatters them; they trust their ability to mold raw athletic talent. I knew I wasn't going to be the guy whom scouts were salivating over. I was relatively undersized for a professional offensive lineman, and I wasn't an athletic freak, the way some of the others were. I also knew that if I blew one of the tests, my chances were probably shot.

I did have a few advantages. One was my work ethic. I had a goal, a few months to achieve it, and a plan—and I had enough confidence by that point to believe that if I stuck to my plan, good things would happen for me. There would certainly be guys who were stronger and faster and bigger than me, but no one was going to outwork me. No one. Another advantage was Jim Ivler, my agent. Jim was a Penn State alum who had worked with several former Penn State players. He had a long face and dark, peaked eyebrows, which gave him a constant expression of enthusiasm and vague surprise—a deceptive impression, because Jim was smart. He knew what was up. I had talked to a few other agents and players about their experiences, and Jim stood out. Long before I'd earned him a dime, I knew he had my back.

Jim arranged for me to train for the combine at the Athletes' Performance Institute in Carlsbad, California, a training center for prospects from around the country. It was a strange environment, down to the palm trees outside my window. Back home, there was snow. Every morning, I would wake up at six a.m. so that I could be at the gym by seven in order to get ready for my first workout—even though it did not start until eight. I needed that full hour for stretching and warming up. After the first grueling day, there wasn't a morning when I didn't need to loosen my muscles and try to rid them of the soreness from the workout the day before.

My strength and conditioning coach, Brent Callaway, and I got along immediately. He saw sprinting, jumping, and agility for what they were: science. *Track and field is just physics in action*, he'd say. So when he wanted me to improve my technique or be more explosive off the ground, he'd talk about angles and vectors. He'd use words such as "levers" and "tension"—and, important to me, he'd use them correctly. He'd watch me sprint, then slowly walk over to me, his arms crossed over his chest, his mouth slightly frowning. *Every movement should maximize your efficiency. That will translate into power and speed*, he'd say, before demonstrating the motion he meant.

After Callaway was done with me, I'd lift weights, bench-pressing hundreds of pounds until failure—the point at which my muscles stopped responding and I couldn't raise the bar from my chest. I also worked with an offensive line coach, Hudson Houck. I could buy a few extra seconds to catch my breath by asking him for stories about winning Super Bowls with the Dallas Cowboys. Then it was time for lunch. Eating food like quinoa took practice too.

THE COMBINE TOOK PLACE in February, in Indianapolis, at the Colts' stadium. Outside it was cold and bleak, but we weren't there to hang out by the pool. Inside it was a circus—335 players, representatives of every NFL team, and nearly a thousand members of the media.

To start, we were lined up in front of hundreds of scouts, wearing nothing but our compression shorts. The scouts looked us over, noting our musculature and the shapes of our frames. They measured the lengths of our arms and the spans of our hands. We were weighed and rotated through medical stations, our blood drawn, our urine tested, our hearts monitored, our bodies poked and prodded as if we were livestock or race horses up for auction. Over the four days of the combine, doctors thoroughly examined us—repeatedly. I had to learn to be patient. Every scar was noticed, every prior tweak discussed, including a few so minor that I had not known about them. I was told about a minor fracture that I had not ever been aware of.

Once we were dressed, we wore shirts marking us with our positions and a tagging number. (I was 48 OL—OL for offensive lineman.) On the third day came the Wonderlic, which was supposed to measure intelligence. The organizers had introduced a new test that year, and I was disappointed not to get a perfect score. The physical tests, though, were what I was worried about. To my relief, I did well, performing in the top ten among offensive linemen for most of the circuit. The only disappointing moment came when we ran the 40-yard sprint. I had managed to keep calm for most of the combine, my nerves usefully humming but not jangled, until I stood at the starting line. Then my breathing deserted me. My heart started racing; I could feel the blood rushing through me. *Come on, John,* I whispered, and I tried to gather myself through breathing exercises. Before I could settle myself, though, I was sprinting, willing myself forward, my arms and legs wildly swinging. No technique.

My time wasn't terrible—just not what I had expected myself to do. Mostly, my pride suffered. I knew the sprint would not affect my place in the draft. The combine is a test of endurance, not only physical endurance and mental acuity but an ability to withstand the attention and the pressure. Teams want to see if you will wilt. They want to find out what will make you crack.

The third part of the combine, along with the medical examinations and the tests, was the interviews. I had prepared for those with the same seriousness that I'd worked on my three-cone drill. I'd memorized the names of all thirty-two offensive line coaches and learned a little about their backgrounds. They were ready for me too, asking me not only about my experiences at Penn State but about growing up in Buffalo. Most of the questions were designed to test my knowledge of football, but there were some oddball questions as well, which was standard in these types of interviews. One team asked me to list everything I could do with a paper clip.

Part of my preparation involved turning every question I could toward football and away from math. In a funny way, I had to be careful. My intelligence was an asset, but only to a point. None of the teams were quizzing me to gauge whether I was smart. My academic record spoke for itself. But at least a few of them were trying to gauge whether I was *too* smart—or not smart in the right ways. The vast majority of football players who are extremely intelligent on the field—who have great football smarts—would have flunked college calculus. Likewise, if you put the top PhD students from Caltech on a football field and had them face a blitz, they would have no idea what to do. (Even if you know football analytics, there's a huge difference between knowing what to do when watching video and assessing all the possible variables and calculating the right decision when a 350-pound nose tackle is bearing down on you. It sounds obvious, but basically every football fan forgets this every Sunday.) To coaches and scouts, my degrees said nothing about how I would perform on an NFL offensive line. It's safe to guess that they might have had another concern when they were deciding whether to draft me. Unlike most guys they were talking to, I had options. *That's a problem*, Jim said.

One of my scouting reports began with the line: "Highly intelligent—will be successful with or without football." While I very much hoped that was true, it was not a point in my favor. In fact, it was more of a warning than praise. The phrase "with or without football" raised red flags. In the NFL, football is supposed to be your life. You are supposed to live and breathe football. You are supposed to give everything you have to the game. And if you don't, there are ten guys standing right behind you who will. *They're going to need to see that you're totally committed to football*, Jim warned me. *Do not—do not— talk too much about math. You're hoping for a long career in the NFL. Stress this: the math thing can wait.*

For some teams, there was another question: Was I worried about my brain?

IT WAS A LEGITIMATE QUESTION. When I was in high school, the reports that playing tackle football could cause lasting brain damage had become front-page news. By the time I graduated from Penn State, the link between the repeated blows to the head and chronic traumatic encephalopathy (CTE), a degenerative brain disease, was well established. For an offensive lineman, the risk was probably even worse, since the cumulative effect of repeated sub-concussive blows— the kind that I experienced dozens of times every full-contact practice and every game—was possibly even more dangerous than getting one concussion. Some of the stories were grim. Former players who committed suicide were autopsied and found to have had CTE. Football players had an increased chance of getting diseases such as Alzheimer's and ALS. Hundreds of former players were suing the NFL.

I knew all of this. No one was hiding anything from me. I knew that I had a lot to lose. People constantly reminded me of it. Not only friends but complete strangers would say to me, with real concern in their voices, *You have such a bright future. Aren't you concerned about concussions?* Others would straight-up call me a fool.

I was honest when I answered. No, I hadn't been brainwashed by the media or hoodwinked by the NFL. No, I wasn't playing for the money. Of course I was aware of the risks. Football is a brutal sport. While there seems to be a lot of individual variation in how the brain responds, it seems pretty obvious to me that repeated blows to the head can't be good for your long-term mental health. So why would I take the risk? The answer was simple: I loved playing football. I loved being part of a team. I loved blocking my opponents. I had a very limited window to play football, and as long as a team would have me, I would do everything I could to earn my spot. I did not try to protect my head when I was on the field. (If I did, that would be the surest way to get seriously injured.) For me, football was an addiction. And it gave me so much more than just a hit of adrenaline. It was part of who I was, and it was what I loved.

The fear of getting a concussion or suffering long-term brain damage didn't keep me up at night. There was a risk, and I didn't know how great the risk was, but I had decided not to worry about it— so I didn't. For better and worse, I had a very powerful ability to train my thoughts on what I wanted to think about and away from what I didn't. I was stubborn too. Maybe I *was* a fool. I certainly wasn't completely rational. But that was the choice I had made, and I was committed to it.

After the combine, I was given a grade of 5.55, which translated to a "chance to become an NFL starter." I was projected to be taken in the third or fourth round. Then, it was really out of my hands.

Twenty-one

John von Neumann

Math, 2014

After the combine, I went back to State College. Done with classes, I wanted to enjoy my final couple of months at the place I'd come to think of as home—though not enjoy it *too* much. I was still in training. I wanted to rerun the 40 at Penn State's pro day, when a handful of scouts would come to campus to see a few prospects who hadn't been invited to Indianapolis and to get a second look at those who had. More important, I wanted to keep my fitness up in case I was drafted at the end of May. If I did have to report for a rookie mini-camp, I needed to be in the best shape of my life. Otherwise, there was no way I would make a final roster.

After my performance at the combine, I knew I had a good chance of being selected, but how good? I wasn't projected to go high, and there are always prospects who fall off all the boards—usually without explanation. Teams devote unbelievable resources to evaluating talent, but at the end of the day it's often a crapshoot. According to the NCAA, there are more than 16,000 college football players eligible for the NFL draft. Only about 250 get drafted. There are always NFL-ready players who don't get picked, and there are players who shouldn't be anywhere near a pro stadium who do. Ranking players, especially given the diversity of positions, not to mention the schemes and situations and experiences that each one had in college, can be just about impossible. No college star, no matter what style his college team played, has ever been in game conditions that replicate the NFL. And no one can predict injuries, burnout, or the random twists and turns of life.

I KNEW THAT MORE and more teams were using statistics and mathematical tools to help them evaluate their picks. There were different approaches. Some teams, like the New England Patriots, stockpiled low draft picks and identified value in the later rounds. Other teams still tried to solve all their problems by drafting a single star quarterback or defensive lineman. Whatever the strategy, though, there were very few teams that were not employing some kind of analytics research and statistical analysis to help them make good decisions. They were trying to see how predictive the size of a player's hands is, or how well speed correlates with game yardage.

For obvious reasons, I thought teams were smart to use analytics and statistics—though I had my doubts that they were always using them correctly. I was far too accustomed to seeing statistical concepts misused, and I knew that sometimes even situations that seemed straightforward were not at all. In one of my classes, I had come across a famous study conducted in 1975 of graduate student admissions at the University of Berkeley, California. The authors of the study,

P. J. Bickel, E. A. Hammel, and J. W. O'Connell, were looking for evidence of gender bias in the admissions process. They found it—or so it seemed. Of the 12,763 applicants for the fall of 1973, about 44 percent of the men were accepted, and just 35 percent of the women were admitted. There were 277 fewer women and 277 more men admitted than would be expected if admissions were gender-blind, assuming that the male and female applicants have similar qualifications. That is a stunning difference.

Then the researchers started delving more deeply into the data. At UC Berkeley, admission for graduate study is generally granted by the faculty of the department that the student is applying to. (An applicant for the PhD program in mathematics, for instance, would be judged by the faculty of the math department.) The authors decided to look at which faculties were most responsible for the bias, and then look to them individually for evidence of discrimination. What they found shocked them. There were 101 departments at Berkeley. Sixteen of those departments turned out either to have no female applicants or to accept all applicants regardless of gender. The authors then turned their attention to the remaining eighty-five departments. They found four with statistically significant signs of bias against women. Those departments accounted for a deficit of twenty-six women. They also found six departments biased in the *opposite* direction, accounting for a deficit of sixty-four men. This was extremely confusing. What happened to the surplus of men? And where were the missing women?

The authors realized that they had run into what is known in statistics as the Simpson's paradox, or spurious correlation. They had assumed that a bias against female applicants would show up as a connection between the gender of the applicant and their acceptance, but they now saw that was a false assumption. The meaningful correlation was between the gender of the applicant and the department to which he or she was applying. It turned out that far more people applied to certain departments than to others. Nearly two-thirds of the applicants to the English department, for instance, were women, while women accounted for only 2 percent of the applicants to the mechanical engineering program. More women were applying to departments where *lots* of people were applying, and fewer were applying to departments that had high acceptance rates. Once the authors of the study took this into account, they discovered that the evidence of bias against female applicants was very small.

"The absence of a demonstrable bias in the graduate admission system does not give grounds for concluding that there must be no bias anywhere else in the educational process or in its culmination in professional activity," the authors cautioned. Bias absolutely exists in academia and the workplace. The authors were showing something much narrower. Further research could look at not why women were being discriminated against in the application process, but why so many women were applying to humanities programs and so few women were applying for degrees in math and science. It would involve questions such as, Why are women less likely to take the prerequisite classes for applying to graduate school in STEM areas? That research would have to take into account influences, biases, educational systems, and cultural pressures from a very young age, beginning in primary school—or probably earlier. (Forty years later, I wonder how much things have changed. There must a reason why, in my math classes, it was more common to see a woman born outside the USA doing well than it was to see an American woman.) It is not enough to just point to statistics and assume they are telling you something clear. You have to ask the right questions to find the right answers.

So I was a little skeptical that all of the teams using analytics to help determine their draft picks were doing so free of faulty assumptions. In football in particular, it can be hard to isolate variables. A lot of scouts end up resorting to their intuition—and this is an instance where intuition can lead people astray. The first quarterback chosen in my draft class, by the Jacksonville Jaguars with the third overall pick, turned out to be a guy from Central Florida named Blake Bortles. Blake was a friend of mine. I'd trained with him in California before the combine. He crushed me at Ping-Pong night after night. I liked him from the moment I met him—as did pretty much every NFL team. His combine and pro-day performances were spectacular. Plus, he *looked* like an NFL quarterback: 6-foot-5, about 230 pounds. Years of experience and probably lots of data told scouts that quarterbacks with Blake's attributes tend to succeed in the NFL. Blake may still become a great NFL quarterback (and I hope he does), but he did not have the start that many expected of him.

Scouts liked to talk about Blake's "prototypical build" and his "stature in the pocket." But given how much money was at play, I'm guessing they weren't influenced by his looks alone. They probably also had analytics to back up their decision to have him high on their draft boards. The Jags in particular had an active analytics department; *ESPN the Magazine* had even written a feature about it the year before. But whatever analytics they used turned out not to be predictive. The point isn't just that evaluating talent is hard, and teams need to be lucky (though that is true). It's that you can have all the data in the world, but it can be meaningless or can be misinterpreted if the right questions and the right contexts aren't considered, and sometimes the right questions aren't the most obvious ones.

I didn't know what kinds of questions teams were asking about me. I just knew that football can be a rough business. It is, after all, a zero-sum game.

I HAD BEEN THINKING ABOUT zero-sum situations since I was a kid—not because I was smart but because I was a kid. Cutting a cake into two parts is zero sum: more for me means less for you. Playing Connect Four is zero sum: if I win, then you lose, and if you win, I lose. Soccer—even if the coaches were pretending not to keep score—is a zero-sum game. Chess is a zero-sum game. A lot of the things I liked to do, in fact, were zero-sum games.

Math didn't involve zero-sum situations. But math helped me understand them. One of the books that Ludmil Zikatanov gave me during my junior year was a book on functional analysis. Reading it, I came across the minimax theorem: a way of determining the best strategy for minimizing the possible loss in a zero-sum game. It made sense to me intuitively. The minimax theorem says that there exists a unique value that represents a gain for one player and a loss for the other such that each player can expect to achieve at least this favorable outcome if they use an optimal strategy. In other words, they can minimize their maximum possible loss.

The basic idea of the theorem and the impressive math behind it immediately intrigued me, and so did the author, John von Neumann—the same man who helped develop the modern computer. In a 1928 paper on strategy in games such as poker, he wanted to figure out how to represent rational behavior under conditions of uncertainty. The work he began then would revolutionize economics. "Real life consists of bluffing, of little tactics of deception, of asking yourself what is the other man going to think I mean to do," he once told the scientist Jacob Bronowski. "And that is what games are about in my theory." There is a psychological element to this, of course, but von Neumann's work did not involve speculation about people's peculiar desires or personality traits. He was not interested in the psychoanalysis and interpretive methods that Sigmund Freud had been working on around the same time in Vienna, not too far from von Neumann's native Hungary. The little tactics of deception that von Neumann referred to did not involve the subconscious. To the contrary, they were explicit. They could be written in the language of reason, the language of math.

Von Neumann had a strong hunch about the importance of his theorem from the start. "As far as I can see," he later wrote, "there could be no theory of games . . . without that theorem. . . . I thought there was nothing worth publishing until the Minimax Theorem was proved."

The more I read about the minimax theorem, the more I was intrigued by its author. Von Neumann quickly became my favorite mathematician. As I read about him and began to explore his work, I had the sense of discovering a figure who stalked the twentieth century, a giant—and now is barely remembered. *How have I not heard of him?* I wondered. *Why did no one tell me?*

The clarity of his thought and the complexity of his work made him a hero to me. His range was astonishing, perhaps unmatched by any modern mathematician. Most mathematicians cultivate their expertise in a single field, or maybe two. Von Neumann on the other hand published well over a hundred papers in his life, making contributions in pure math, applied math, physics, computer science, biology, even weather forecasting. During World War II, he perfected the implosion method for exploding nuclear fuel, a crucial advance in the development of the atom bomb. The famous concept of mutually assured destruction is a consequence of his work in game theory. He made advances in artificial intelligence and is even credited with first describing a self-reproducing computer program, now known as a computer virus. "Whatever field he touched was profoundly affected by his thought," his collaborator Oskar Morgenstern wrote after von Neumann died at only 53 years old, from cancer.

Von Neumann was born in 1903, in Hungary. When Eugene Wigner, a Hungarian Nobel Prize winner and von Neumann's childhood friend, was asked how twentieth-century Hungary had produced so many geniuses, Wigner replied that it produced only one genius: von Neumann. There are countless stories about how quick his mind was. He moved to Princeton in 1933 to accept an appointment at the Institute of Advanced Studies, right around the time of the rise of Hitler— fortunately, since he was Jewish. His office at the IAS was next to Einstein's, whom he would annoy by playing German march music obnoxiously loud. Why is he so much less famous than his neighbor? He was just as outsized a character as Einstein, known for his well-tailored suits, his parties, his off-color stories, his largeness in every sense. His wife, Klara, one of the first computer programmers, is said to have joked that he could count everything but calories. He liked cocktail parties, power, pretty women, and his Cadillac, which he had a habit of crashing.

Von Neumann's genius lay in taking underdeveloped concepts and brilliantly expanding, improving, or transforming them. Few people have been behind so many revolutions. His impact on the world that we live in today cannot be overstated. Every American kid, I believe, should know his name.

One of von Neumann's most intriguing ideas, to me, is his conception of mathematics itself. He balanced the theoretical and the practical, the ideal and the real. He also listened to internal, or what he called "aesthetic," considerations. He probed the ways mathematics was related to the natural sciences, and the ways in which it was different, moving between theoretical and experimental disciplines. He saw how mathematical ideas grow out of real practical questions, but as they develop, "aesthetical motivations" predominate. In that sense, mathematics is like a creative discipline. But when mathematics loses its connection to empirical sciences, he believed, there needed to be a "re-injection" of empirical ideas. What stayed with me—what I keep in my mind, always—is that tension between the creative and the descriptive. Von Neumann never limited himself, one colleague wrote in a remembrance. "He wasn't afraid of anything." He was the kind of mathematician I aspired to be.

And as I faced the draft, I kept that in mind. I would not be afraid. I would not let one thing define me—including the draft.

But that didn't mean I wanted it any less.

||||||||||||||||||||||||||||||||||||||||||||||||||||||||||||||||||

Twenty-two

Becoming a Raven

Football, 2014

The draft began on May 8, 2014. I was able to relax and watch the first round on TV like a fan, knowing there was no chance my name would be called that day. The second day of the draft—rounds two and three—also came and went without a phone call from my agent, Jim, to tell me that I had been picked. By the third day, I was starting to feel nervous.

It did not help that cameras were following me. I was one of five prospects that the NFL Network had picked to shadow during the draft for the show *Hey Rookie: Welcome to the NFL*, which aired on ESPN2. I was never comfortable when cameras were following me, even in the least stressful of circumstances. But I took it as an opportunity to push back against the stereotype that college football players never cared about school, and to make math seem cool to kids. There was also a new consideration. I wasn't going to play for the money, but that didn't mean that I didn't want to make some. I wanted to help my mom out, given all that she had sacrificed for me, and I suspected that sooner rather than later I'd be a graduate student who would be otherwise broke. The exposure would help—but during those few days during the draft, it was excruciating. With every round that passed, the cameras seemed to zoom in on me a little closer.

My mom came up to State College from the Baltimore area, where she had moved two years earlier, for the draft weekend. We spent the day playing board games, with my phone by my side. I sat with my back to the television. My mom had cooked—fried chicken, macaroni and cheese, baked-bean casserole, sweet potatoes, hot dogs, apple pie—and friends stopped by throughout the day. (Maybe they were there to support me—but they were also there for the food. My mom's mac and cheese was famous.) The fourth round came and went. No call. The fifth round moved along a little too quickly for my liking. When I heard that my hometown team, the Buffalo Bills, was on the clock, I looked up from a game of Settlers of Catan, a little hope in my heart. It stung when they drafted a guard—my position— and it was someone else.

Is your phone working? my mom asked.

Yeah, my phone is working! I was stressed.

The 173rd pick was made. Then the 174th. I had been projected to go in the third or fourth round, and here we were, nearing the end of the fifth. I was starting to wonder whether I would be drafted at all.

Two picks before the Baltimore Ravens were set to make their selection, my mom's phone rang. She looked at it. *That's funny*, she said. *I don't recognize the number.* She answered it. *It's for you*, she said, and handed it to me.

Apparently, my phone *wasn't* working. Fortunately, I had given my mom's as a backup. It was Ozzie Newsome, the Ravens' general manager. He politely introduced himself, and then asked me a question: *Why do you want to play football?*

For a second, I was tongue-tied. It sounded so innocent, the kind of question I had answered a thousand times. But I knew how many questions and doubts were behind it. *How committed to football are you?* he was silently asking. *Why should we use a valuable draft pick on you if you're going to give us only half of who you are?*

The clock was ticking on the television screen. Newsome was waiting for me to speak.

I cleared my throat, wondering how to answer. I hadn't expected this call—not seconds before I might be selected. I think it's safe to guess that no other player in that draft was asked it. At the same time, it was a fair question. A draft pick is a huge investment. If the Ravens were going to bet on me, they deserved to know what I would risk for them.

Football is my passion, I said. *I just love the game. Playing in the NFL has been my dream since I was a little kid.* The clichés poured right out of my mouth.

Newsome paused. *You know, we have a mutual acquaintance. Donovan Smith*—one of my Penn State teammates, who was from Baltimore—*and I use the same barber.*

Really?! I said, trying to inject as much enthusiasm into my voice as possible while figuring out what to say. I couldn't believe that I was on the phone with the general manager of the Baltimore Ravens, during the draft, talking about a barber. *What a funny coincidence. Donovan is a good guy. Great player. Great, great guy.*

Newsome chuckled. I could tell he was having some fun with me.

Finally, he said, *Well John, we're going to draft you.*

I thanked him, and then he passed the call along to the coaches and logistics coordinator, and after what seemed like both an eternity and no time at all, I hung up. Relief washed over me, followed by joy. A few seconds later, I heard my name announced on TV. *Thank you, Jesus,* my mom said. It was official. I'd made it to the NFL.

I PACKED UP MY HONDA Fit that night. I needed to be in Baltimore the next morning. Rookies were expected to report immediately. About a month after I got to Baltimore, during the training period that precedes camp—known as "organized team activities," or OTAs—one of the guys at the Ravens' facility asked for my car registration, so that he could set up access to the player parking lot. *My registration is expired,* I said.

Well, you've got to get it registered immediately, he replied.

I can't, I said. *It won't pass the inspection.* A Jimmy John's delivery guy had sheared off my front bumper in a hit-and-run, so I'd gone to Home Depot, bought a drill and some industrial ties, and reattached the bumper myself.

He stared at me and shook his head. *You need a new car,* he said.

But this one drives just fine! I protested.

You need a new car, he said again, and sighed—a sigh that seemed to say, *What have I done to deserve this?*

Jim, my agent, was only too happy to oblige once I got back to State College, in the break between OTAs and the start of training camp. He had connections everywhere. What did I want? A truck? A Benz? As it happened, there was a car in State College that I had my eye on, I replied: a used 2013 Nissan Versa hatchback.

You're kidding, Jim said, then tried again. *There's an Audi dealership too. I can get you a great deal.*

I want the Versa, I insisted.

How about a new one? Jim countered.

Why would I want a new one? This one only has 30,000 miles!

So I drove down to Baltimore as the proud owner of a 2013 black Nissan Versa hatchback. When my new teammates saw me, they did not let me live it down. *What the hell is that?* They saw it in the parking lot, tucked between a tricked-out Suburban and a giant Ford truck. *Can you even fit in there?*

I don't know, Ursch, Jeremy Zuttah, the starting center, said to me, shaking his head. *I have a bad feeling about this. It's not safe. If I hit you on the highway in my truck, I'd kill you.* He even looked up the Versa's safety ratings.

But to me, the car was perfect. Great gas mileage, no trouble scooting into tight parking spots, and, best of all, a good price: about $9,000, after I traded in my Fit.

Even my dad, one of the stingiest people I know, thought I was crazy. After all, I'd just signed a four-year, $2.364 million contract. But I had done the math. There was a high probability that I wasn't actually going to see all that money. My contract was not guaranteed. That $9,000 was about 6 percent of my $144,560 signing bonus—and that signing bonus might have to last me a very long time. I was convinced that I wouldn't see the rest of the money, because I wouldn't make the team.

ONCE THE VETERANS arrived for off-season training activities, the O-line room was uncomfortably full of interior linemen. The starters were already set: Jeremy Zuttah at center, Kelechi Osemele at left guard, and Marshal Yanda, a future Hall of Famer, at right guard, the position I had played in college. Several of the players I was competing with for one of the few backup spots had solid experience as starters. Others were former draft picks. I didn't think I had much of a shot.

Reggie Stephens, a practice squad player, took me under his wing—but carefully. *When I was a rookie, there was a vet who seemed like he wanted to help me out,* he explained. *He gave me a lot of advice, but it turned out the things he told me were wrong. He was trying to sabotage me. So I'm here for you—but I want to show I'm doing it in a pure way. You have a question, you come to me.* I did. He gave me tips on my technique and how to be in the league. He taught me how to command an offensive line when playing center, and how to make my way around the Ravens building—whom to befriend, whom to avoid, what to do, when to speak. So did A. Q. Shipley, a former Penn Stater—even though it was clear that we were fighting for the same spot on the roster. Not many guys would have done that. And sure enough, while I made the final roster, Shipley was cut. I knew that was the way business was done, but it still shook me.

For some reason, our offensive line coach, Juan Castillo, took a special interest in me. The other guys joked that he treated me like his son. I think he felt we understood each other. His parents, Mexican immigrants, had been incredibly hardworking. His dad, a shrimper, woke up at 3:30 every morning to go to work. He drowned when Juan was eleven, and his mother worked two jobs, one as a maid and another busing dishes at a restaurant, to support the family. Juan had his parents' work ethic. It helped him become the first person in his family to graduate from high school. It helped him turn a one-semester scholarship into a starting linebacker position. After he became a coach, he contacted the teams with the best coaches—the Dallas Cowboys, the University of Notre Dame, the San Francisco 49ers—and drove all over the country to observe and learn from them, spending the nights in his car.

When he was working for the Philadelphia Eagles, he once told me, the head coach, Andy Reid, recommended Malcolm Gladwell's *Outliers* to him. *I'm not much for books, Coach,* Juan answered.

But you'll like this one. It's about how you need 10,000 hours of practice to become great at something, Reid said.

Aw, shiiiit, Coach! I already know that, Juan replied. *I need to have a talk with the author. He stole that idea from me.*

There are different kinds of coaches: strategists, intimidators, commanders, delegators, teachers. Juan was a great teacher. He had a soft drawl, and there was something conspiratorial to his tone. He looked at my stance—square to the line, flat back—and the way I moved toward the ball, and saw how it might be improved. *You want your back at more of a slant,* he'd say. He'd crouch, bringing his compact build low to the earth, and show me where to put my feet, how to come off the line. If I was going to have a good shot at backing up the interior of the line, then I needed to be more versatile, so he worked with me at center. *You could have a long career at center,* he'd tell me. *Make some good money.* These weren't the kinds of changes you could make in a single day. We would meet before practice to work, and stay after practice to work more. He was only asking of us what he demanded of himself. Whenever I went to the Ravens' facility, whatever time of day (or morning or night), I knew he was somewhere in the building.

I WAS LUCKY. Not everyone is. I knew that the NFL would be different from Penn State, but I didn't know just how different it would be. At Penn State, there was a closeness to the team that came almost automatically. The jersey itself was its own strong bond. In the NFL, we were mercenaries. I quickly made a best friend on the team, a rookie left tackle from the University of North Carolina, James Hurst.

Hurst was taller than me, with close-cropped blond hair and a beard he could grow like a lumberjack. He faced even tougher odds of making the team than I did. After breaking his leg in the bowl game of his senior year, he had not been drafted. He had signed with the Ravens as a rookie free agent. But he was a good player—and, more important to me, a good guy. We'd watch horror movies or play video games, or I'd go over to his place and do math while he and his girlfriend watched TV, and it almost felt like college. But that was the exception. For the most part, after practice, guys went their separate ways.

An NFL locker room is very different from a college one. Players had arrived by a trade or a signing, and within a few years they'd leave that way too. Most of them did not live in Baltimore full time. They had permanent homes (and sometimes families) in Iowa or Miami or New Jersey. We were all in different stages in our lives. Some guys were twenty-one; some guys were thirty-three. Some guys couldn't wait to get home to see their wives and three kids, to whom they were devoted. Some guys were juggling three girlfriends. Some guys had their houses equipped with generators and enough canned food to survive the end of civilization, which they believed could come soon. Some guys spent a lot of time, without any attention or fanfare, volunteering in the community. The kind of distinctions that divided American society but seemed totally irrelevant in a college locker room—like race and religion—were a little more visible in an NFL locker room. Still, there was rarely any tension. We called one another brothers, and on game days we would feel it, but mostly we were colleagues. Even when the video tape of Ray Rice, our running back, hitting his fiancée in an elevator captured the nation's attention, the locker room was calm. It affected us as men, of course; it was deeply troubling. But it didn't shake the team as a whole in the way it probably would have at

Penn State. Most guys tended to mind their own business. Football was our job, the same way accounting was an accountant's job, or medicine was a doctor's job. Whether or not we loved football, we referred to practice as *work*, which is what it was.

There was also a competitive element to the dynamic. We were all teammates, but we were also gunning for a limited number of spots. At the end of training camp, in early September, half of us would be cut. In college, we had competed with one another for playing time and pride. In the NFL, we were also competing for paychecks. Even if you weren't mean-spirited, you had to look out for yourself. Careers are short and contracts not guaranteed. When you are fired, there is no notice or severance. If another team does not pick you up, you are out of luck—and a lot of players are out of money. Some guys don't have a college degree, and some of those who do have degrees that are basically worthless. Even the guys who make millions a year sometimes have trouble making ends meet after their careers. It goes fast when you've got a couple of fancy cars in the garage, when you spend a few hundred at dinner and then a few thousand at a club without blinking. I knew what I was doing when I bought that used Versa. Even so, it can be hard to imagine a future without football, when football is part of your identity.

In itself, the competition wasn't a bad thing—in fact, the opposite. Competition has driven me since I was a little kid. It's the way I motivate myself, test myself, prove myself. But there are dark sides to it, obviously, especially in the workplace, and not only because not everyone plays fair. Before I came into the League, I took it for granted that businesses should be free to act in their self-interest, but the more I saw of a system that was so heavily tilted against players, against my teammates, the more I wondered whether that was right.

I WAS RELIEVED TO MAKE the team that September. I spent the first few weeks of the 2014 season inactive, on the sideline wearing street clothes. (Not all players on the roster dress for every game.) Then in early October, Kelechi Osemele, our left guard, injured his knee. When we headed down to Tampa, I knew I'd be in pads—and unless Kelechi had some kind of miraculous recovery, I would be starting in his place.

I was anxious all week. If I had a bad game, I might never get another shot. To complicate my situation, I was playing on the left side of the line for the first time in my life. Everything was backward—my hands, my feet. I had spent a decade training my body to move on instinct. My hand would punch and my foot would plant without my thinking about it—only now that was the wrong hand, the wrong foot. If I hesitated or pivoted the wrong way, I knew I would pay for it. In all likelihood, I'd spend some time going up against All-Pro defensive tackle Gerald McCoy, who had been tearing up offenses that season. *He's in his contract year*, Marshal Yanda, our right guard, warned me. *Make it your business to know when guys are playing for extensions. That's when you gotta worry.* (Two weeks later, McCoy would sign a seven-year extension for nearly a hundred million dollars.)

Mostly, though, I worried about the heat. The forecast in Tampa called for humidity and temperatures in the high eighties. I hated playing in the heat. Generally, I could ignore the weather altogether when it was cold. I had played in sub-freezing temperatures, the wind whipping off Lake Erie, in short sleeves and thought nothing of it. I could ignore the rain, and snow falling so thick it accumulated on my shoulders. The heat was something else. It was insidious, a life-draining force. All week, I checked the weather obsessively, hoping that some front would move in and spare me. No such luck.

That Sunday, I stood on the sideline in front of a giant fan. I had an IV before the game to give myself some extra hydration, but even before kickoff, I felt the sweat gathering. It was a relief when I had to run onto the field. As soon as the huddle broke, my vision sharpened. The sound of the crowd, roaring to distract us, grew dull and distant. I lowered myself into my stance and forgot about the heat.

Two minutes into the game, I caught sight of the Tampa safety streaking toward our running back, Justin Forsett, who was thundering through a hole I'd created. I dove, taking the safety's feet out from under him, and springing Forsett, who shot down the field for a 52-yard gain. I stood up, simultaneously amped and settled. I was ready. The Ravens ran behind me twelve times for 86 yards that game, and the line shut down McCoy: no sacks or quarterback hits. We scored 38 points in the first half alone on our way to a 48–17 victory.

After the game, I didn't let myself get too excited. I thought of what OB, my old Penn State coach, would have said: no single victory is everything; no loss is a reason for despair. *Be happy tonight, but there are things you could have done better.* I would study tape of the game the next day.

On the plane back to Baltimore, I passed out from exhaustion. After Hurst and I grabbed some Panda Express on the way back from the airport, I planned on going straight to sleep. My body had never felt so beat up. My hand was throbbing where a nose tackle had stepped on it; my knee cracked when I bent it; my head had a dull ache. But when I lay in bed, I was too wired from the game—not to mention all the caffeine, equivalent to ten cups of coffee, that I had taken right before the game. So I got up, found a pen and a clean piece of paper, and focused on an unproved conjecture. Immediately, I felt myself calm down.

I STARTED AGAINST the Atlanta Falcons the following week, and then moved back to the sideline when KO recovered, getting a few snaps here and there but usually treating Sundays as a time to watch and learn. For the most part, that meant studying Yanda as closely as I could.

Yanda didn't look that special off the field. He had a bushy blond beard and the kind of solid build that suggests cornfields and bales of hay. (Sure enough, he grew up on a farm in Iowa.) But on the field, his quick feet, good technique, and canniness were immediately obvious. In one game late in my rookie season, I watched him help our right tackle double-team the defensive end with a quick punch, slide over to handle the nose tackle, and then block the inside linebacker. I had never seen a player handle three different defenders quite like that. It took a combination of awareness, quickness, strength, agility, and instinctive calculation that no one else could match. Of course, to him, it was no big deal.

I couldn't hope to copy Yanda's style. For one thing, he did a few things unconventionally, giving space where other guys were aggressive, daring defenders to bull-rush him. It worked—but only for him. With everything else, though, I tried to do what he did. For one thing, he understood the importance of the cohesion of the line. Yanda always did his assignment. If he was supposed to nudge, he just nudged. If he was double-teaming, he double-teamed. He didn't get overly aggressive or try to do too much, which could screw up the spacing and angles for everyone else. He didn't call attention to himself by doing something showy. Part of his power was in his self-control and restraint. Plus, he was tough. *Really* tough. If toughness is currency in the NFL, then Yanda is up there with the richest men in the league. The stories about his toughness are legendary. Before I got there, he had himself tasered to win a bet. I once was standing on the sideline with him during a game when he mentioned that his foot was bothering him. He pulled off his shoe and then stripped off his sock, which was drenched in blood. The callus on the ball of his foot had come off. He calmly pulled off the hanging skin, put his sock and shoe back on, and ran back onto the field. His thumb stuck out at an odd angle. A few seasons later, he would tear his rotator cuff. Instead of ending his season and getting surgery, as any other player would have done, he simply moved to the other side of the line—and he was *still* the best offensive lineman in the league. He did not think any of this was particularly heroic. He was just doing his job, doing it as he thought it was supposed to be done. That's what made the biggest impression on me.

TOWARD THE END OF THE SEASON, our right tackle went down with an injury in the second half of a game against the Texans. I ran onto the field to take over for him. It didn't matter that I had not played right tackle since high school. I was a professional now. It was my job to figure it out.

You're playing right guard, Yanda said when I got to the huddle.

No, I insisted. *Juan told me right tackle.*

No way. You're going to get enough J. J. Watt as it is. I hesitated, wary of contradicting my coach—but then took my place at right guard. I was even more wary of contradicting Yanda. Besides, Yanda was right. I didn't want to get killed. J. J. Watt was the best defensive player in football, and that season you could make a case that he was the most valuable player—offense or defense—of the entire league. The previous season, he'd recorded 20.5 sacks, and he looked likely to match it this season. That stat doesn't even begin to capture his impact on opposing offenses. He was scary. A YouTube video of him doing a 57-inch box jump and then squatting 700 pounds once made the rounds. No one was more intimidating. Against us, he had been having, as usual, a great game. We ended up losing the game, and I can't say that I came away feeling like I had bested J. J. Watt, but I did play well enough to gain confidence going into the playoffs. I would be starting the rest of the way.

We played our rivals, the Steelers, in the first round, at Pittsburgh. As I stood on the field, waiting for the coin toss, I looked around. It was a freezing night, made worse by a bitter wind, but I could actually feel the angry, rabid heat radiating from the crowd. The stands looked like a swirling mass, as the fans waved their Terrible Towels, and the air was filled with the raw sound of people screaming and singing "Renegade." When we took the field, all of that fury was directed at our tiny huddle. I relished it for a moment, loving it, before putting my head down and tuning it out.

The Steelers had blown us out the previous time we'd played, in the regular season. This time, it was our turn. The most beautiful sound I have ever heard in my life was the sound of the silence near the end of the game. We had shut the crowd down.

The following week, we traveled to face the New England Patriots. I was playing as well as I ever had. Everything I'd been working on was coming together. Halfway through the game against the Patriots, I was playing the best game of my life. It didn't matter that I was up against Vince Wilfork, a five-time Pro Bowl player who was listed at 325 pounds but looked like he weighed about 50 pounds more. I not only handled him but was dominating. The team was clicking on all fronts. Everything was going the Ravens' way. Well into the third quarter, we were leading 28–14.

With the defense on the field, I sat back and watched the game on the jumbotron. Then I noticed something strange was happening. The referee announced that one of the Patriots was an "ineligible" receiver. The announcement came again—and again. On the field, the defense was obviously confused. Yanda turned to me, bewildered. I knew that if Yanda hadn't seen it before, then something really strange was happening.

On one play, it looked like the left tackle ran straight down the field and caught a quick pass for a first down. If the receiver *was* in fact the left tackle, that would have been illegal. For each play, the offense has to have seven players on the line of scrimmage, and only the player on each end of the line is allowed to catch a forward pass. But the Patriots were using an unbalanced line, with four offensive linemen instead of five. The player who lined up where the left tackle would normally be wasn't actually a left tackle. He was on the end of the line, so he was an eligible receiver. On the right side of the line, there was a receiver in the slot, the one who had reported as ineligible. He was on the line of scrimmage and "covered" by another receiver to his right, so it would be illegal for him to catch a forward pass. The defense was as confused as we were. No one knew whom he was supposed to cover.

The Patriots' offense, which hadn't clicked all day, was moving efficiently down the field. Tom Brady, the Patriots' quarterback, found the player who looked like the left tackle for a first down, then the slot receiver lined up next to the "left tackle" for another first down, and then the "left tackle" again for yet another. At this point, our coach, John Harbaugh, ran onto the field, screaming, protesting that we weren't being given enough time to identify the ineligible receiver and adjust. He took an unsportsmanlike-conduct penalty to make his point. I looked across the field at the Patriots' head coach, Bill Belichick. He was standing on the sideline, the hood of his navy Patriots sweatshirt hiding the pom-pom of his winter hat, his face as unreadable as rock.

The Patriots scored a touchdown on the drive, then never used the formation again. It wouldn't have worked if he had; we'd have been able to adjust. But its effects were lasting. We were rattled. They scored again with another trick play, tying the game with under five minutes in the third. They would go on to beat us and then would go on to win the Super Bowl.

Trick plays happen all the time in football. Sometimes they work; sometimes they don't. Usually they're a little cheesy. I spent a lot of time thinking about that one, though. Its success depended on the surprise, but Belichick wasn't pulling it out just to gain a few cheap yards. It demonstrated a deeper understanding of what was happening on the field. Early in the game, the Patriots lost their center to a knee injury, wrecking their already weak offensive line. Tom Brady's quick release and accurate passing were keeping his team in the game, but barely. He was sacked twice, and the Patriots were able to run for only 14 yards on 13 carries in the entire game. Their backup right guard was having an especially rough game. We knew they weren't going to be running the ball much, which gave us a real advantage. We knew where to attack the line, and we could more or less plan for a pass. But Belichick took that away. Just as important—maybe more so—he got in our heads. He changed the dynamics of the game. I thought back to what von Neumann had said about game theory, that it consists of bluffing and anticipating the actions of others.

There is no chance that Belichick was thinking of von Neumann when the Patriots called out that play. Still, I suspect that he is familiar with von Neumann's game theory. Belichick, after all, had been an economics major at Wesleyan. He is brilliant at reading the situation and applying an optimal strategy guided by intuition honed by decades of experience and the information he has at hand. "There are so many factors in football that it's really hard to find two situations that are the same," Belichick told reporters not long ago. "Even in some situations that are similar there's usually something in there, the conditions on the field, or the game, or the wind, or something else that adds another variable in there besides just point-differential and time and timeouts. But I would say even with the three timeouts involved, which could be three, two, one, zero, so there's another four possibilities there and the field position and, again, the score spread, the differential in points. . . . When you put it all together, again, we have our guidelines and there's certainly a feel, if you will, for certain things." A *feel*. He was talking about intuition—and he was doing it in a way that sounded familiar to me. I wasn't fooled by his tendency to scoff at relying too much on analytics. He thinks like a mathematician.

‖‖

Twenty-three

Challenging Conventions

Math, 2015

After the season was over, I headed back to State College. I had nothing holding me in Baltimore. Even the furniture in my apartment had been rented. State College was still the place where I felt most at home. I would do my off-season training there—working out twice a day, usually kickboxing during one session and lifting during the other—while continuing to do math with Ludmil and my other collaborators at Penn State. The math department had even created a position for me as an adjunct researcher. I was also eager to embark on an ambitious project of my own, my first major solo paper, involving something called centroidal Voronoi tessellations (CVTs). A CVT is a special type of Voronoi diagram, or geometrical structure in which a certain distribution of generating points is optimally partitioned. It's useful for doing things like clustering and data compression—problems that are increasingly relevant to the modern computer-driven world.

There is a famous historical example that neatly explains a Voronoi diagram. It's from the nineteenth century, actually before Georgy Voronoi, after whom the diagrams were named, was even born. In 1854, a cholera outbreak swept through Soho, London, killing 10 percent of the population within days. Doctors assumed that the disease was spread through the air, and they had no real idea of how to stop it. A physician named John Snow was convinced that it spread through contaminated water, not the air, but no one would listen to him. Then he made a map. On his map, Snow represented deaths in each household with little black bars, which showed how the deaths clustered. The map was, in fact, a graph. Snow had a hunch that the contaminated water came from a particular pump, on Broad Street. But given that there were other pumps nearby, how could he prove it?

His solution was to partition the map into separate areas, drawing a curve along the line where it would take equal time to walk to neighboring water pumps. This is now known as a Voronoi diagram. Voronoi diagrams are essentially models for the distribution of data, where each subsection, or cell, contains one point (called the generator) such that every point within the cell is closer to its own generator than to any other generator. In Snow's Voronoi diagram—a term that would not be invented for another fifty years—the water pumps were the generating points and the borders of the cells were defined by the length of time it took to walk to them. Those who lived within the curve around the Broad Street pump used it as their source of water. And sure enough, as Snow showed, most of the deaths occurred within that area. Establishing the link between water and cholera would revolutionize both sanitation systems and medical research, saving countless lives.

In mathematics, sometimes the urgency of a real-life situation helps lead to new breakthroughs; sometimes theoretical results turn out to have applications that we never could have dreamed of. John Snow never could have anticipated that Voronoi tessellations would have applications in, say, computer graphics. That's often how mathematical discoveries work. When you use Google to search for something on the internet, Google's PageRank algorithm uses something called Markov chains, which generate very random outcomes dependent on initial conditions, to determine the importance of a web page. Markov chains grew out of the work of Andrei Markov in the late nineteenth and early twentieth centuries, long before computers, let alone Google, existed. When Markov himself looked for applications to his chains, he could have no idea that someday they would be used to model real-life processes such as electron-behavior algorithms and identifying genes. Instead, Markov used his processes to study alliteration in Alexander Pushkin's novel-in-verse *Eugene Onegin*. We can't always see how things will fit together, or where chains of events will lead.

I TOOK THAT TO HEART. Still, that didn't mean that I would just wait for things to unfold. That has never been my nature. That spring, 2015, my decision to put off pursuing my PhD until after I was done playing professional football started to eat away at me. I had always prided myself on not sacrificing football for math or math for football, but if I was being honest with myself, I had prioritized football. I was not pushing myself to the limits of my potential in math. As I worked on the CVT paper, I realized that I did not even have a good sense of what my limits were.

There was a lot that I could do on my own in mathematical research. I did not lack for self-motivation. But I had so much more to learn from other mathematicians. I could pretend that I had all the time in the world to accomplish everything I wanted to when I was done with football, but that simply was not true. If I was serious about becoming a good mathematician, then I needed to become serious about pursuing that goal while I was still young.

I had been able to balance coursework and football in college. Why, I wondered, did that seem impossible now? To be sure, the game in the NFL was more demanding, and I had a responsibility to the people who were paying me. Still, I saw opportunities that I wasn't taking. There was a long off-season break, in which coaches were not even allowed to contact players. Most guys went back home or traveled. Other guys were going back to school, still pursuing their undergraduate degrees after declaring early for the draft. In fact, the NFL encouraged it. Why couldn't I do the same—only for a PhD?

Certainly, coordinating my football schedule with a PhD program would be harder than it had been at Penn State. Most obviously, I would not be able to be on campus during the fall semester. But mathematicians are problem solvers. Why, I wondered, couldn't we figure out a solution?

I started researching different programs. The math department at the Massachusetts Institute of Technology stood out for its mix of theoretical and applied mathematics. Being a student there would give me the chance to do a lot of theoretical computer science, too, which I wanted to study, along with dozens of other areas. The flexible but exacting structure of the program suited me. Plus, some of the best mathematicians in the world were on its faculty. MIT was the only program that I ended up applying to.

Just before the start of training camp, I learned that I had been accepted—and that MIT would allow me to begin in the spring semester of 2016.

‖‖

Twenty-four

Concussion

Football, 2015

A couple of weeks into the start of training camp, in early August, during a full-pads practice, I pulled right to trap out the outside linebacker, Terrell Suggs, who was barreling in my direction. He slammed into me. The next thing I knew, I was looking up at a crowd of concerned faces—coaches' and trainers' faces. I could hear Suggs's voice somewhere in the distance.

My memory of what happened next is disjointed. I was later told that a cart drove onto the field to take me to the locker room, but that I shouted and cursed at the coaches and refused to ride it. I have no memory of that, though it sounds like me. There was no way I was going to be driven off as long as I could physically walk. I do remember that when I got to the locker room, I realized that I was crying.

The NFL-mandated test to see whether I had a concussion was pointless. There was no question that I did.

By then the pain in my head was becoming overwhelming. I could call it a headache, but it was unlike any headache I had ever had. The pain was at once acute and diffuse, concentrated in my skull but also somehow present all around me, radiating from the lights overhead, throbbing in every sound I heard. *No screens*, the doctor told me. That meant no computer, no TV. *I'd take a break from reading until you can do it without pain,* he added. Standing for more than a couple of minutes made me nauseated. The sun was like a hammer to my head.

A fireman who worked with the Ravens drove me back to the hotel where the team stayed during training camp. I went into my room and closed the blinds. Then, woozy, I sat down. What could I do? Lie in bed with the lights out. For how long? There was no timeline.

Every morning, I went to the facility. Some days, I'd take a test that measured my cognitive function against the baseline score that I had achieved when I was healthy. I failed it three times in a row. I did a little stretching and rode the exercise bike or ran on a treadmill until the pain in my head became overwhelming. I usually lasted a few minutes. Eventually, I could make it for a mile. Then, while the rest of the team practiced, I would go back to the hotel. Unable to read or watch TV, I tried listening to audiobooks. I tried writing a little. I tried playing chess against the computer. I tried doing some math. That was a mistake. I couldn't do math.

The thing about a brain injury is that you can't look inside the skull to see the extent of the damage. I just had to wait. Slowly I started to get better. Still, even when the pain subsided, everything felt off. I didn't have any idea when I would be well. After about three weeks, I was given a different cognitive test than the one I had kept failing. This one, I passed.

You're going to see the independent neurologist tomorrow, I was told when I went into the facility a couple of weeks after the concussion. The way the statement was phrased made me startle. *The* independent neurologist? Not *an* independent neurologist?

I went. Aside from my balance, which wasn't great, he thought I was well enough to play. He did bring up the question of whether I *should* play, though. He was a football fan—he got tickets to Ravens games twice a year, he told me. (Any more would violate the terms of his "independence.") But he was also a neurologist. He knew what the risks of football were, and he knew that I stood to lose more than most.

If you're interested, when you get to Boston in the spring, we can do a brain scan, he offered. Doctors couldn't yet diagnose any lasting damage to the brain so soon after a concussion, but they were making strides. I thanked him and said I probably would, but secretly I knew I wouldn't take him up on the offer. The truth was, I did not really want to know.

I had always assumed that if I had a major concussion, one that knocked me out cold, that would be enough to make me quit football. In fact, I'd assumed that was the *only* thing that could make me stop, short of a debilitating injury to my body. Strangely enough, the opposite happened. It only made me more determined to play. My fears had come true, and I had survived. It hadn't been so bad, I told myself. One of the ways the brain protects itself is by forgetting.

Having made the decision to keep playing, I did what I had always done. I shoved the questions and doubts to the back of my mind, far from my daily thoughts. I knew the risks. If I did have CTE, then a protein called tau would accumulate in cells throughout my brain. Over time, it could lead to memory loss, cognitive problems, and behavioral changes. But I didn't know whether I had CTE or would develop it. I didn't know what the long-term effects of playing football would be on my brain. At this point, I *couldn't* know with any certainty.

Headlines about concussions and football sometimes made it seem like every NFL player was bound to get CTE, but I read the studies. I knew there was a lot of variance from player to player. There are people who haven't had any concussions who get it, and people who have had many concussions who get it. There are also people who have had severe concussions who don't. So-called subconcussive hits may be more directly associated with the development of CTE than a full-blown concussion, and the duration of a career, not the number of concussions, may be correlated best with the disease. Even so, I wanted to keep playing. I made the choice to live and play with that uncertainty. The chance to keep playing football at the highest level was worth it to me.

Even after I was cleared to play, I wasn't fully myself. I was quick to anger. A stray, meaningless comment could make me feel aggressive, make me want to fight. I had panic attacks. Anxiety would grip my chest, like a vise. After a few hours, or occasionally a few days, it would fade. Little things would trigger it—and one big thing. I couldn't do high-level math.

My CVT project was way beyond what my brain could handle at the time. That was as painful as anything I experienced physically. I was so close to finishing the paper, so close to producing one of the biggest results of my career—and suddenly, all the ways forward were closed off to me. Even easier mathematics seemed too hard for me. I'd reach for a theorem that I *knew* I knew, and it wouldn't be there. I would try to visualize patterns, or to stretch or twist shapes—a skill that had always come particularly easy to me—and I would be unable to see the structures or make things move. I was confident that my skill would come back. But I didn't know whether there would be lingering effects. A concussion is a brain injury, after all. Some guys tear their ACLs and come back at full strength. Others are never quite as explosive. How could I know how my brain would rebound?

Eventually, after a few months, my brain became quick and flexible again. I reached for definitions, and there they were. I recognized patterns and was able to manipulate concepts. I finished the CVT paper, which was accepted by one of the top journals in the field. There were no lasting effects. On the field too it was as if the concussion had never happened. When our center went out with a torn labrum, I took over his starting spot and, aside from a few shaky shotgun snaps, thrived in it. On the field, I was not afraid at all. I didn't hesitate for a moment when I crashed into a defensive tackle. On the field, I never once thought about trying to protect my head.

Twenty-five

Uncertainty

Math, 2015

Many people have wondered how I could take so much risk and live with so much uncertainty every time I stepped onto the field. Part of it was my personality, my ability to compartmentalize and to keep unwelcome thoughts from creeping in. Perhaps it was also because I am comfortable with the concept of uncertainty in a way most people aren't. As a mathematician, I confront uncertainty all the time. It is there, half hidden, in probability theory. It is part of machine learning and the dynamical systems I explored in celestial mechanics. It is in the question of what a proof can prove, and what constitutes truth.

I thought about uncertainty a lot. Uncertainty suggests that there are not only limits to what we know, but limits to what we *can* know. Discovering why this isn't true is not why I became interested in math, but it is part of the story. I got a glimpse of it taking physics in high school, but it wasn't until I started reading a book that Xiaozhe gave me about quantum mechanics, when I was getting my master's, that I started to understand how profound and challenging the math behind the field is. The classical mechanics developed by Newton explained much of the universe—the part we could see—extremely well, but the laws of classical mechanics broke down when it came to atoms and subatomic particles. For many years, physicists (and mathematicians) were puzzled. The better they were able to measure the behavior of things like electrons and light, the more they were confused. Sometimes it seemed as if nature, at the smallest scale, was made up of waves. Sometimes it behaved as if it were made up of microscopic individual particles. Sometimes it seemed continuous and sometimes discrete. Sometimes it behaved as if it were both.

Things only got stranger from there. When physicists tried to measure the movement of subatomic particles, no one could say with perfect accuracy where a particle was at any single instant, or in what direction it was traveling. If they knew what forces were acting on it, then they should have been able to predict where it would be and how it would be moving in the future (at least in theory). In fact, that had been the motivating idea behind physics for three hundred years. But with subatomic particles, that information wasn't enough. The conventional model for an atom, for instance, involved a large central nucleus like the sun, with little electrons, following fixed orbits, whizzing around it like planets. But if the subatomic particles followed classical laws, then the electrons would spiral into the nucleus. Clearly, atoms don't collapse that way. Classical laws, which were supposed to be universal and absolute, don't hold.

Then, in the 1920s, the young German physicist Werner Heisenberg did something strange: instead of seeing uncertainty as a problem to be solved, he devised a fundamentally new subatomic theory which, among other things, implied a form of uncertainty. Heisenberg had an insight about the intuition physicists were using to describe atoms: they were trying to visualize them, to come up with a picture—in this case, the solar system. That model was familiar, so it seemed intuitive. But instead of describing the inner workings of an atom with a picture, Heisenberg tried to describe them in the language of math. After all, no one had ever actually *seen* inside an atom. They had only measured how it behaved. In a series of papers, he, Max Born, and Pascual Jordan worked out the mathematical formulation of quantum mechanics. Heisenberg represented the observed properties of electrons as matrix calculus.

Matrices lack a special property. They do not necessarily commute. By this, I mean that when we are dealing with regular algebra, we can safely say that $xy - yx = 0$, where x and y are variables each representing some arbitrarily chosen constant. It does not matter what order you put them in. But when we are dealing with two matrices, the order in which they are multiplied *does* matter. Why? It has to do with eigenspaces and eigenvectors.

When we have a particular kind of matrix, the kind that Heisenberg was using, the matrices only commute if they have the same set of eigenvectors. If they do not share eigenvectors, then the order in which they are given matters. If the matrix represents the physical property of an electron, the set of eigenvalues is the set of all possible values that an observed quantity can have. If you measure a physical property—say, the position—then you get an eigenvalue, and the corresponding eigenvector is the state of the system at that moment. If two observable properties don't have any eigenvectors in common, they cannot occupy the same state at the same time, which means they can't be measured precisely at the same time. The problem isn't with the tools we use or with the experiments that we design. We will never be able to develop a technology that allows us to predict perfectly certain pairs of related properties. It is a fundamental feature of quantum systems.

In 1927, Heisenberg published a paper titled *Über den anschaulichen Inhalt der quantentheoretischen Kinematik und Mechanik*, or "On the visualizable content of quantum theoretical kinematics and mechanics." The *anschaulichen* was his great concern here—the visualizable, the perceptible, the intelligible, the intuitive. He wanted to account for the consequences of matrix mechanics. What he found was that the more precisely we know the position of a particle, the less precisely we know its momentum. This is known as the uncertainty principle.

It's important to realize what the uncertainty principle does *not* mean, especially since it has been so abused by popular culture. It does not mean that we cannot know anything precisely. It doesn't mean that we can't trust what we observe, or that life is an illusion. Life may be an illusion (see, for instance, Nick Bostrom's 2003 paper published in *Philosophical Quarterly*, "Are You Living in a Computer Simulation?"), but that's not what the uncertainty principle says. What it does say is that we can't determine certain pairs of related variables, like position and momentum, precisely at the same time. It also tells us that until a state is observed, we cannot talk about properties, only probabilities. Sometimes it can tell us that we'll know something with high probability. Sometimes it spreads out the odds—there's a 6 percent chance here, a 4 percent chance there. When it comes to certain complementary properties, the more confident we are that we know one, the less confident we are that we know the other.

Many people—including many great physicists—were horrified by Heisenberg's work and its implications. Erwin Schrödinger, an Austrian physicist, saw that Heisenberg's quantum mechanics was so weird that he found it "repellent." How could a particle exist in multiple states until it is observed? In a letter to Einstein, he proposed a thought experiment: Imagine a cat in a sealed box, he said, with a radioactive atom and a flask of poison. When a sensor detects the decay of the atom, it causes the flask to break, releasing the poison and killing the cat. Until you open the box (or, perhaps, hear a scuffle or meowing), you do not know whether the cat is dead or alive. But here's the twist, Schrödinger argued: until you open the box and make your observation, the cat is simultaneously dead *and* alive. Schrödinger came up with a whole new formulation of an electron's behavior by approaching an electron as a wave instead of a particle—and it turned out to be equivalent to Heisenberg's picture.

Even now, there are very different arguments over how to interpret quantum mechanics. Maybe the future and the past interact in a transactional way, like a handshake. Maybe all possible futures are real; maybe we live in a multiverse. Maybe no one really does understand quantum mechanics completely, even though its predictions have been successful.

Given how easy it is to torture the uncertainty principle when you turn it into a loose and careless analogy, I'm wary of doing it myself. And yet, it's tempting. So often, people want to divide the world into two. Matter and energy. Wave and particle. Athlete and mathematician. Why can't something (or someone) be both?

AT THE END OF JANUARY 2016, I moved into a short-term sublet in Boston, just over the Charles River from MIT. The best thing about the apartment was the long hallway, perfect for practicing shotgun snaps. (Or almost perfect. The apartment was filled with oil paintings and antiques, which were practically begging to be strafed.) The morning after I arrived, I went to campus for the first time. I walked through the columns of the grand classical main entrance and found my way past the engineering labs and lecture halls to the building that housed the math department. It was like stepping into my personal vision of paradise. Chalkboards lined the walls of the hallways in the math building, and people stopped to use them. Casual conversations quickly became discussions of open conjectures. I passed the small offices of some of the greatest mathematicians in the world, noticing the extra chairs across the desk or in the corners—chairs where a visitor, such as myself, might take a seat to discuss a thorny problem.

I started reading the major papers of some of the professors. When one of them seemed particularly interesting, or when there was a conjecture that sparked an idea, or when I came across a concept or proof that I did not quite understand, I would stop by the author's office to discuss it. One of the papers was on something called the max-cut problem, by Michel Goemans. As I read it, I marveled at its elegance. It was ambitious in its reach, thorough in its execution, clearly and stylishly written. I not only admired its brilliance but was also inspired by it. Professor Goemans is one of the leading experts in combinatorial optimization, a field in which I had no experience. My background was more in graph theory and numerical analysis. Still, as soon as I read the max-cut paper, I knew that I wanted to learn from him and work with him.

When I entered Professor Goemans's office to talk about his paper with him, he peered at me through his round, horn-rimmed glasses a little skeptically. I expected this. Mathematicians, after all, take nothing on faith—and as the only professional football player at MIT, perhaps I had a little more to prove than others. I needed to show that I wasn't a novelty. After we spoke a little about his paper, he gave me a new problem to think about. *Don't work too hard on it*, he said with a light accent. *You probably won't solve it.*

I started to work on it in my head as soon as I stepped out of his office. I thought about it through dinner and until I fell asleep. I thought about it on the bus to campus the next morning and throughout the day. I didn't stop thinking about it until I had it figured out.

I went back to his office a few days later with the solution. *I want you to be my adviser,* I said at the end of the meeting.

You should probably talk to several other professors first, he replied. *It's a big decision.*

You're going to be my adviser, if you'll have me, I said. He looked amused.

Fine, he said.

Twenty-six

‖‖

MIT Football

Football, 2016

MIT even had a football team.

On Monday mornings, I joined the football team for practice. Granted, they weren't NFL players. At 310 pounds, I probably weighed about fifty or sixty pounds more than the biggest guy on MIT's O-line. But when we ran, they put me to shame. They could outsprint me.

The MIT football program had changed since my mother tried to get the Engineers to recruit me when I was a senior in high school. The team used to be so small that it had trouble finding enough players to play both offense and defense. Sometimes there were no more than forty guys at practice. But in 2013, the Engineers started recruiting for the first time in their history. It wasn't easy. Since the program is Division III, the coaches couldn't promise admission, and the applications from recruits are held to the same rigorous academic standards as the rest of the student body. Nevertheless, MIT went 6–3 in 2013, and 10–1 the next season—the first back-to-back winning campaigns in the team's history. Seventeen players on the 2014 team were high school valedictorians, and the starting quarterback was majoring in aerospace engineering—otherwise known as rocket science. (Sorry, Mom.)

At MIT, even for the football team, academics still came first. At Penn State, meetings were mandatory, workouts could not be missed, and practice time was sacrosanct. We planned our coursework around our football schedule. There were rewards for that kind of discipline, of course: the status that came with being a football player, the possibility of playing in the pros, the sound of a hundred thousand people roaring.

At MIT, on the other hand, most practices were in the morning before classes began or during the schoolwide activities window from five to seven p.m. (MIT actually sets aside time for students to stop studying, to encourage them to do extracurriculars or simply take a break.) If a player had to miss practice because of class, or if he showed up late because he was busy with his schoolwork, there was no punishment and there were no questions asked. The weight room was about one-twentieth the size of the weight room in the football building at Penn State—and it was for all thirty-three varsity sports, not only football. For much of the day, it was empty and shut. The strength coach, a great guy named Tim Viall, was also the football team's offensive coordinator and offensive line coach—as well as being the strength coach for every MIT team.

I didn't know what to expect when I showed up at practice that first Monday. But what I found was that the team at MIT is no joke. It is a real football team. In some ways, there's something even more pure about the way MIT plays than any team I'd ever been on. These guys were there because they wanted to be. No one was making them come to practice; no one was checking up on them. They knew that football is dangerous. They knew the feelings of exhaustion and pain. They still showed up. They didn't do it for money, and they didn't do it for status. The average size of their crowds was fewer than a thousand. On campus, no one gave them a second look. (The guys who won the Putnam Competition—a national math contest—three years in a row were the ones who got treated like star quarterbacks.) But they came every day and worked hard because it was their choice—because they loved to play football.

When I needed a break, I played chess. I had continued playing chess casually online ever since picking it up again in college, but I was becoming more serious about it. I started ordering chess books on opening theory and studying famous games. I carried around Jesus de la Villa's *100 Endgames You Must Know* and referred to it so frequently that the pages started falling out. I started thinking about trying to become a national chess master—an ambitious but, it seemed to me, achievable goal, if I set myself to it. Chess, with its dual emphases on calculation and intuition, appealed to me naturally. I could spend hours thinking about a single position.

In April, I packed up my sublet and headed back to the Hyatt Place outside Baltimore, where I would live for seven weeks during off-season training activities. MIT's semester was not over—there was still a month left—but I could send in my coursework by email. The first week of workouts with the Ravens was optional—but not, I knew, for me. I had to show them that the football team, not school, was still my priority.

I said goodbye to MIT reluctantly. I refused to clean out my cubicle in the first years' shared office space, leaving behind a crumpled sweatshirt and backpack in the corner, and books scattered across my desk. I wanted it to look like I was only out for a cup of coffee, not gone for good. I told my good friend and officemate, Jake, to rearrange my stuff periodically, so that it would look like I was still there. I knew I wasn't fooling anyone. The only person I was really trying to deceive was myself.

Around midnight before I left, I went running on the path along the Charles River, and looked at the dark outline of MIT's main campus buildings across the water. As I ran, I could only think of how I did not want to leave.

JUST BEFORE LEAVING BOSTON for Baltimore, I learned that Kelechi Osemele, our left tackle, had signed a big contract with the Oakland Raiders. Juan, the Ravens' offensive line coach, let me know that they were planning on moving me to the left side of the line to take his place. I had only ever played a few games on the left side, at any level of football. It wasn't really a big deal; I was a quick study and a hard worker, and I trusted myself to make the change without any problems. Still, overriding instincts that have been trained over a decade is not easy. It can be unsettling.

When I got back to Baltimore, I started wondering whether I still loved the game in quite the same way that those MIT players did—the way that I had in college. Football had always been worth the time it required. It had been worth the risks to my body and my brain. I had never wondered whether I wanted to keep doing it. Nothing had seemed more important. I was starting to question whether that was still the case.

I wasn't thinking of quitting football. I still felt a visceral thrill when I did my job well. Nothing else made me feel so alive, so close to contact with something elemental within my heart. But now there were other things I loved—and I was starting to think about whether I owed them more time and attention than being an NFL player allowed. After off-season training was over, I went back to State College for about six weeks. I was training intensely there, kickboxing and lifting, but I was also working with Ludmil and coauthoring a few papers with a few MIT professors. My projects were beginning to multiply.

Once training camp began at the end of July, I shoved all this aside, determined to focus on the job that paid the bills. Then, early during training camp, during a preseason game against the Carolina Panthers in early August, I felt something in my shoulder sharply yank. I kept playing and finished the game, but that night, after the adrenaline wore off, the pain was too much to ignore. I had separated my AC joint. It wasn't a serious injury—after a few weeks of rest, the trainers built a kind of scaffolding to protect my shoulder so that I could start training again, and after a while, I could play at full strength without feeling it too much. My shoulder was fine. Even so, I was struggling more than usual, on the field and off.

More and more that season, I was looking forward to Sunday evening, after the game was done. As soon as I got home from the stadium or airport, I would unwind. I would grab a math book or my clipboard and some blank paper. I worked on a few papers and took two classes that fall, sending in my problem sets by email. I limited my work to our off day and the bye week. I knew that it would have seemed like a distraction—and I understood why. It was fair for my coaches to expect me to focus only on football, given what they had invested in me. It was hard for them to understand that doing math actually relaxed and rejuvenated me. It helped prepare me for the long and punishing week. Besides, I did not have a choice: MIT does not allow PhD students to be enrolled only during the spring semesters. The math department was willing to take the rare step of letting me take classes via correspondence, but I had to be enrolled as a full-time student throughout the year. And the truth was, I wouldn't have wanted it otherwise.

For most of the week, though, my attention was consumed by football—both preparing and, increasingly, recovering. My body was starting to feel the cumulative effects of more than a decade in the sport. Already my fingers were crooked, and my thumb was starting to bend alarmingly like Yanda's. My knees clicked when I bent my legs, my neck snapped when I moved my head a certain way, and my heart was under some strain from carrying the extra weight required of offensive linemen. I made it back onto the field after my shoulder healed and started a few games, but my body felt wrecked. In a game against the Jets, in October, I had ended up on the bottom of the pile of several goal-line pushes—with more than a thousand pounds on top of me. On the train back to Baltimore, I could barely sit.

I could handle the pain. It certainly wasn't enough to make me want to stop playing. There were other issues, though, that I found increasingly hard to ignore. That season, I was starting to confront my limitations as an NFL player. It didn't matter how much effort I exerted or how much I trained. At practice and during games, the gap between what I wanted my body to do and what it could do was harder and harder to close. When I had been a rookie, coaches had been forgiving of my flaws because they believed they could fix them. I had potential. But in my third year of the game, I was starting to think that my level just was what it was. At some point, working my ass off wasn't going to be enough. Some guys are irrationally optimistic about their talent. It probably helps them. But I tried to be honest with myself. I was a good football player, but I did not have the body size or the explosive raw athleticism to be a truly great guard. I did think that I could have become a decent starter and have had a long career as a center, but I had a realistic sense of how high my ceiling was.

I kept these thoughts to myself, of course. When I came to the facility, I worked as hard as if I were a rookie with dreams of becoming a Pro Bowler. But I knew that wasn't the path I was on. I spent the last few games of the season on the bench, as the team made a playoff push. I tried to enjoy it—to listen to the crowd, to appreciate the small rituals that had defined so many days for me over the past decade: loosening my muscles, taping my hands. I started talking with my family and a couple of my closest friends about how much longer my career would last. I had the sense that it wouldn't be forever.

We played our second-to-last game of the season on Christmas Day, against our rival, the Pittsburgh Steelers. It had the atmosphere of a playoff game—which, in a way, it was. If we won, we were in the postseason; a loss would keep us out. With 14 seconds left to play, we were up 27–24. It looked like we would make it. Then I watched in disbelief as Antonio Brown, a Steelers receiver, pulled in a catch and blew by three Ravens defenders to score a touchdown. And just like that, our season was effectively over.

The mood in the locker room afterward was grim. Losing always hurts, but this one was more painful than most. I felt it too. The short flight back to Baltimore was quiet. There wasn't anything to say.

But when I got home, I felt a guilty kind of relief. Making the playoffs would have extended the season into January. And that would have been a problem for me, because in the first week of February, I was supposed to take my PhD qualifying exams.

Twenty-seven

Passing a Test

Math, 2017

Most students spend an entire semester preparing for their quali-fying exams. It is no easy task: three professors grill you for two to three hours on three areas of mathematics in front of a blackboard. If they mention a theorem, you are not only supposed to know what it is, but you are also expected to pick up the chalk and write the proof. There are hundreds of lemmas, conjectures, definitions, and examples to remember. More important, you're supposed to be able to discuss the material in an intelligent way, and to be able to synthesize it, connect it, and interpret it.

The failure rate is decently high. If you fail once, you get a second chance within a certain window of time. If you fail it again, you're kicked out of the PhD program.

I had only that January to prepare for it.

I moved the large blackboard that normally stood along the back wall of the bedroom into the small study, where it blocked the door. I could barely get past it into the room—which meant that I could barely get out. Essentially, I trapped myself in there. I gathered my books and scoured them for any theorem I might be asked about, any conjecture that could be relevant, and made exhaustively comprehensive study sheets. Every night, I would stand at the blackboard, chalk in hand, and test myself, as if it were the real thing. It was the part of math I liked least—all that memorizing, all that recitation. But I knew that it was necessary, and that having immediate access to those ideas in my mind would quicken and strengthen my own ideas.

I left the blackboard only for a quick visit to Buffalo to see my dad and a whirlwind trip to Houston during the first week of February for an event before the Super Bowl. The contrast between the lavish parties in Houston and my monkish lifestyle at home while studying gave me whiplash.

I took the exam the Wednesday after I arrived in Cambridge. I was nervous—more nervous than I had ever been before a football game. It wasn't that I lacked confidence in my grasp of the material. I had it down cold. But given the scope of what they could ask me about, I couldn't discount the possibility that something unexpected would come up. It felt like my future as a mathematician was on the line.

At the end of the exam, I was told that I had passed. I left the room feeling a little lighter. It meant I was able to start a new stage in my degree, focusing more on research. Mostly done with coursework, I spent that semester collaborating on a couple of papers with a few MIT professors, playing around with a few open conjectures, and sketching out possible projects. I was on track to finish my PhD in a few years—or even sooner, if I wanted. Before too long, I could become a professor.

I played a lot of chess that spring. I started to play it almost obsessively. I'd play it on my computer or even on my phone, starting a blitz game if I had a spare three minutes. Sometimes I simply analyzed positions in my head. Most of the time there was a chess instructional video on YouTube or a stream of a major tournament on the TV screen. While I worked or relaxed, analysis of the Catalan opening was on as background noise. I started studying the former world champion Vladimir Kramnik's games with the same intensity that I had studied Jake Long's football games at Michigan as a kid. At night, I would fall asleep to the soothing voice of Yasser Seirawan, a grandmaster and one of the best chess analysts, as he explained a position on TV.

I met a grandmaster, Robert Hess, at a charity event, and he quickly became one of my best friends. We could spend hours talking about anything and everything—life, Chipotle, chess lines. I thought more seriously—and more urgently—about my ambition of becoming a titled chess player. I wasn't content to wait a few more years before I started training. I wanted to do it now.

I was happy in Cambridge, living a short walk from MIT. Around midnight, I'd often go running. In mid-April, I was due back in Baltimore for off-season training, and I needed to be in shape. But the closer my return drew, the further I felt from it.

II

Moving On

Math Meets Football

I flew down to Baltimore in mid-April. It was even harder to leave Cambridge than it had been before. There was a lot on my mind— and not only math. Earlier that spring, my fiancée and I decided to try for our first child. A few days after I got back to Baltimore, I learned that, if all went according to plan, I would be a father by Christmas. I welcomed the news—but it also made me think. I had never worried about getting hurt or what kind of long-term damage football might cause to my body before. Now I realized that I cared about my longevity. I wanted to be in good shape for my daughter as she grew up.

The thoughts about moving on to the next stage of my life that had been in the recesses of my mind, out of the range of my everyday thought, started pressing to the front. I talked more with the people closest to me about my future in football and out. I told them that I had one more year on my contract with the Ravens, and my impulse was to see it through. I didn't like the idea of breaking an agreement, and I wanted to retire on my terms. But of course there was more to it than that, even if it was hard for me to talk about. Football was, and had been for almost as long as I could remember, part of my identity.

Whatever questions I had about my future disappeared whenever I was inside the Ravens' facility or on the field. I didn't let my effort slacken. I was competing for the starting center job, and I was determined to win it. I trained like my job was at stake—and I felt like it was. If I didn't get the spot, I wasn't sure I'd make the team at all. During the monthlong break between minicamp in June and the start of training camp at the end of July, I stayed in Baltimore so that I could keep working out and training at the Ravens' facility. After lifting, I would spend extra sessions there working on my snaps and technique. I was completely focused on preparing for a good season. But when I got home, the doubts would creep up.

The day before I was supposed to head to the facility for physicals and meetings at the start of camp, I saw an article in the *New York Times*. It was about the publication of a study that examined the brains of 111 deceased NFL players and found that 110 showed evidence of CTE. Nothing in the study surprised me, or told me anything new, or changed my understanding of how playing football impacts the brain. In fact, I was irritated by the way the study's findings were reported by some media outlets, leaving the impression that 99 percent of NFL players have CTE. My friends in the league started texting me to see what I thought. *Listen*, I said, *that's not what the study says. There's not a 99 percent chance that you have CTE*. I couldn't say for sure, but it was extremely likely that there was a strong self-selection bias in the sample. The brains that were examined came from a donation bank. Those brains were there because either the deceased or their families suspected the player had CTE. They were not chosen randomly from a general population of NFL players. I was not about to tell any player that he should retire. Everyone had to make the decision that was best for himself.

The question of what was best for me, though, was hard to answer. I stared at the images of the brains that lined the sides of the *Times* story. While the study did not change the way I felt about football, it did force me to examine my choices. How could I best serve the talents I was born with? How could I best serve my family and society? What was I willing to risk? What did I hope to gain? I thought about the opportunity I had to reach kids who might otherwise turn away from math—especially African-American kids, who had too few examples of black mathematicians. I have never in my life felt that the color of my skin has affected my math, nor how I have viewed myself as being perceived, but I knew not everyone was so lucky. While tutoring or doing outreach, I had sometimes met with young African-American would-be mathematicians, and listened to them ask how I've managed to get to where I am. I watched them hold back tears when talking about being behind or feeling like they can't succeed because they don't have the background that the "elite" students in their classes have. I felt a responsibility to show them by my example that they could succeed.

I had maybe two or three years left in the PhD program at MIT. What did I want to do with them? What kind of mathematician could I become if I gave it my full focus? Both football and math rewarded youth. Which did I want to spend these valuable years on?

That afternoon, I continued the dialogue with my friends, my family and with myself. For my whole life, football had been the thing I cared about most. Now, I realized, there were other things I cared about more.

Still feeling unsettled, I packed my bags that night to head to camp. I would play one more year, I told myself. Just one more. The next day, I checked into the team hotel, went to my physical, and reported for meetings. Practice was set to start the next day.

All the while, the uncertainty was growing. I was having a hard time imagining myself on the field instead of at a blackboard. That night, sitting in my hotel room, I talked more with my loved ones on the phone. *I'm ready to call it,* I finally said. I talked to my agent and a couple of friends. *Did you get to decide the length of your contract?* a friend asked me.

No, I conceded.

Then why should that dictate what you do?

The next morning, I called John Harbaugh and told him that I would be retiring. *Is there anything I can do to convince you to stay?* Harbaugh asked.

No, I answered. I told him that I loved football, and that I had nothing but respect for him and the Ravens, but it was the right moment for me to devote myself to my studies at MIT. We wished each other the best.

THAT FALL, for the first time in more than a decade, my weekends were my own. Two months after retiring, in late September, I flew from Boston to Chicago, and then took the train down to Macomb, Illinois, a small town surrounded by farmland, where Ty was the offensive line coach at Western Illinois University.

That Saturday, the Leathernecks played South Dakota State University, a matchup of two of the top Division 1-AA teams. I spent the cool, sunny morning sitting in a folding chair at a tailgate, eating wings and chips, and marveling at the strangeness of the scene. I had been to a tailgate before a football game only a couple of times since

I was a little kid. I spent the first half of the game on the sideline and the second half sitting in the stands. That also felt odd.

Sitting in the stands gave me a good view—not only of the action on the field, where I focused on the offensive line (partly out of habit), but also of Ty. I watched as he paced along the sideline, wearing a headset, his hands on his hips. He had lost sixty pounds since we played together, and instead of having long hair, his head was now shaved. Still, he looked completely familiar to me—not so much as my friend but as a football coach. I had seen the same anxious, impatient posture, the same stalking walk, the same gestures of encouragement and irritation, at every level I had played. Even from a distance, it was obvious that he was a good coach—and that he cared. He seemed to live and die with every play.

One evening back at his house, he handed me a beer and settled into the couch.

So, how's the math going? he asked. I told him about the papers I was putting together for my PhD thesis, and about the conjectures I was hoping to prove. *I think I've got a good approach. Right now it's mostly a hunch, something my intuition is telling me could work. I give it about a five percent chance of working.*

That sounds pretty low, he said.

Nah, man. For something like this, it's actually high.

You need some help? I solved that problem for you that one time, he said.

What problem? I was genuinely confused.

You know, the one with the thing—you know, when we were in the hotel.

I started to laugh, thinking of Ty's claiming credit for the spectral bisection paper I had done with Ludmil. *No way, man.*

Ty told me about his players—where they had come from, where they might be going, why they made him care about them and about the team. *Are you more nervous during games now or when you were playing?* I asked.

I'm way more nervous now. It's not even close, he said. *At Penn State, I cared about winning, but mostly I just had to concern myself with the way I played. Now I've got to worry about all my guys.* He quickly shook his head. *It's not even close,* he said again.

We talked about the schemes that Western Illinois ran, told old stories from college, and caught up on news of our friends.

So, what're you gonna do with that degree when you're done? Ty asked. *You gonna get a fancy job?*

Gonna be a professor, I said.

He nodded.

We talked about my daughter—Ty's goddaughter—who was due to be born in two and a half months.

What if she doesn't like math? he asked. *What if she just wants to run around outside?*

I can find math in that! I said, and he laughed. *The key is that she won't even know she's doing math. She'll think she's playing games. She'll just want to win. What kid doesn't love games?*

I took a sip of beer. I thought of the hours I had spent as a child doing puzzles, long before I had imagined a career. I thought of the thrill I had felt when I started to play football, and when I was tested on the field. I wanted my daughter to challenge herself, and to find fulfillment wherever her interests were. Looking over at Ty, I thought of my teachers and teammates, my mentors and friends. *Honestly, I don't care if she becomes a mathematician,* I said. *I just want her to grow as a person, and to learn how to think.*

Acknowledgments

With special thanks to Glenn Carson, Megan Fu, Ty Howle, Xiaozhe Hu, Venita Parker, Evan Thomas, Oscie Thomas, John D. Urschel, Jake Wellens, and Ludmil Zikatanov.

We are also indebted to our editor, Ann Godoff; to our agents, Sarah Chalfant, Rebecca Nagel, and Steve Ross; and to Casey Denis, William Heyward, and everyone at Penguin Press.

A number of books and sources were helpful references in writing this book, especially: *Fourth and Long,* by John Bacon; *How Not to Be Wrong: The Power of Mathematical Thinking,* by Jordan Ellenberg; *Einstein's Clocks and Poincaré's Maps: Empires of Time,* by Peter Galison; *Incompleteness: The Proof and Paradox of Kurt Gödel,* by Rebecca Goldstein; "The Legend of John von Neumann," by P.R. Halmos in *The American Mathematical Monthly*; *A Mathematician's Apology,* by G.H. Hardy; *Alan Turing: The Enigma,* by Andrew Hodges; *The Ghost Map: The Story of London's Most Terrifying Epidemic—and How It Changed Science, Cities, and the Modern World,* by Steven Johnson;

ACKNOWLEDGMENTS

"From Parlor Games to Social Science: von Neumann, Morgenstern, and the Creation of Game Theory 1928–1944," by Robert J. Leonard in *Journal of Economic Literature*; *John von Neumann*, by Norman MacRae; obituary of John von Neumann by Oskar Morgenstern in *The Economic Journal*; *The Martian's Daughter*, by Marina von Neumann Whitman; "Intuition and Logic in Mathematics," by Henri Poincaré (trans. G.B. Halsted); and *The Making of the Atomic Bomb*, by Richard Rhodes. *The Stanford Encyclopedia of Philosophy* (plato .stanford.edu), *Wikipedia*, and *WolframMathWorld* were also useful starting points.

For those interested in further reading, we recommend: Ellenberg's *How Not to Be Wrong*; Hardy's *A Mathematician's Apology*; *The Man Who Loved Only Numbers*, by Paul Hoffman; *How to Solve It*, by George Pólya; *The Joy of X: A Guided Tour of Mathematics from One to Infinity*, by Steven Strogatz; and any puzzle collection by Martin Gardner.